PROTOPLASMATOLOGIA
HANDBUCH DER PROTOPLASMAFORSCHUNG

BEGRÜNDET VON

L. V. HEILBRUNN · F. WEBER
PHILADELPHIA   GRAZ

HERAUSGEGEBEN VON

M. ALFERT · H. BAUER · C. V. HARDING
BERKELEY   TÜBINGEN   NEW YORK

MITHERAUSGEBER

W. H. ARISZ-GRONINGEN · J. BRACHET-BRUXELLES · H. G. CALLAN-ST. ANDREWS
R. COLLANDER-HELSINKI · K. DAN-TOKYO · E. FAURÉ-FREMIET-PARIS
A. FREY-WYSSLING-ZÜRICH · L. GEITLER-WIEN · K. HÖFLER-WIEN
M. H. JACOBS-PHILADELPHIA · N. KAMIYA-OSAKA · D. MAZIA-BERKELEY
W. MENKE-KÖLN · A. MONROY-PALERMO · A. PISCHINGER-WIEN
J. RUNNSTRÖM-STOCKHOLM · W. J. SCHMIDT-GIESSEN

BAND V

2

THE NUCLEAR MEMBRANE
AND NUCLEOCYTOPLASMIC INTERCHANGE

WIEN

SPRINGER-VERLAG

1964

THE NUCLEAR MEMBRANE AND NUCLEOCYTOPLASMIC INTERCHANGE

BY

C. M. FELDHERR
EDMONTON

J. G. GALL
MINNEAPOLIS

L. GOLDSTEIN
PHILADELPHIA

C. V. HARDING
NEW YORK

W. R. LOEWENSTEIN
NEW YORK

A. E. MIRSKY
NEW YORK

WITH 32 FIGURES

WIEN
SPRINGER-VERLAG
1964

ISBN-13: 978-3-211-80690-6   e-ISBN-13: 978-3-7091-5476-2
DOI: 10.1007/978-3-7091-5476-2

The Nuclear Membrane and Nucleocytoplasmic Interchange

Preface

By

ALFRED E. MIRSKY

Rockefeller Institute, New York

The essays in this little volume give the reader a good idea of what is now known about the structure of the nuclear membrane and of how materials pass through it. These essays form a coherent whole, although no attempt has been made to fit them together smoothly. As for other problems of cell biology, investigation of the nuclear membrane is prospering because many different experimental procedures are being used. The contributions to this volume are an example of how well the different approaches complement each other.

It is known, of course, that both big and small molecules pass through the nuclear membrane. Innumerable tracer experiments demonstrate that many kinds of small molecules penetrate the nuclear membrane. It has long been presumed by cytogeneticists that large molecules can pass through the nuclear membrane, for if a gene product, which moves from the chromosomes out into the cytoplasm, is to possess the specifity of the gene, the product must consist of big molecules. Now that it is known that "messenger RNA" is a gene product, it is well established that macromolecules pass through the nuclear membrane into the cytoplasm. The protein described by LESTER GOLDSTEIN in this volume provides an example of the passage of a protein through the nuclear membrane—from the cytoplasm into the nucleus.

Is passage of both large and small molecules through the nuclear membrane simply a process of free diffusion? Are the pores in the membrane simply holes—and on a molecular scale quite big holes? This seems unlikely, although there are electron microscopists who refer to "the obvious continuity between nucleoplasm and cytoplasm via the pores." Control is an important factor in biological systems. A membrane that is no more than a coarse sieve would imply that interactions between nucleus and cytoplasm are subject to very little control.

Careful observations with the electronmicroscope, such as those of AFZELIUS and WATSON, have in fact shown that the pores are not simply

holes. WATSON (1959) proposed that these structures be called p o r e
c o m p l e x e s, and he then went on to say, "The high degree of structural
organization imperfectly suggested by our observations leads us to emphasize our earlier conclusion that any passage of materials which may
take place through the pore complex is probably controlled and not a
random process."

Much of what we now know about the nuclear membrane points to
the conclusion that passage of materials through it is under somewhat the
same kinds of control as passage through the cell membrane. Many small
molecules do not penetrate the cell membrane by free diffusion but are
transported through the membrane by an active, enzymatically promoted
process. Macromolecules, when they do enter the cell, are brought in by
pinocytosis. The experimental evidence that active transport and a process
resembling in some respects pinocytosis occur at the nuclear membrane
is far from complete. This evidence, much of which is given in the present
volume, will now be briefly summarized.

1. The penetration of amino acids into isolated thymus nuclei is a
sodium-dependent "transport" or "gate" process having many attributes of
an enzymatic reaction (ALLFREY et al. 1961). Competition studies indicate
that specific receptor sites are concerned with the transport of different
amino acids. These sites can readily discriminate between the D- and
L-isomers of the appropriate amino acids but close structural analogues
such as L-isoleucine or glycine can competitively occupy the sites involved
in valine or alanine transport, respectively. After obtaining this information about transport of amino acids across the nuclear membrane, the
same experiments were done on amino acid penetration into a number of
cells, nucleated and non-nucleated erythrocytes and the frog ovum (IZAWA
et al.). The experiments showed that in these cells transport of amino
acids into the cell has essentially the same characteristics as transport
into isolated thymus nuclei. Penetration of amino acids into either the cell
or nucleus is not a simple diffusion process.

2. The potential difference across the nuclear membrane and the electrical resistance of the membrane can be measured by inserting a microelectrode into the nucleus in much the same way as the electrical properties
of the cell membrane have been measured by inserting a microelectrode
into the cytoplasm. In an elegant study by WERNER LOEWENSTEIN (chapter 2
of this volume) these measurements have been made for the nuclei of the
salivary cells of *Drosophila* larvae. The measurements demonstrate de-
cisively that in these nuclei the membrane is a diffusion barrier; the so-
called pores do not offer direct contact between nucleoplasm and cytoplasm.

3. A well known characteristic of the cell membrane is that proteins
(of 65,000 molecular weight and over) do not freely diffuse through it into
the cell; nor do such proteins readily diffuse through the nuclear membrane.
In a convincing series of experiments (chapter 4 of this volume) FELDHERR
and HARDING injected proteins into the cytoplasm of oocytes and showed
that the nuclear membrane prevented their diffusion into the nucleus,

although the proteins (serum albumin and gamma globulin) have a diameter much smaller than the so-called pores.

4. Electrical measurements of the nuclear membrane, experiments on the transport of amino acids into the nucleus and observations on the non-penetration into the nucleus of certain proteins, all show that the nuclear membrane is a diffusion barrier. What, then, is the significance of the pore complexes in the nuclear membrane? In a series of well conceived experiments on the ameba and on the frog oocyte (chapter 4) FELDHERR has shown that coated gold particles do not diffuse freely through the pores but that they do pass through the pores into nuclei. His experiments, still incomplete, point to a process somewhat akin to pinocytosis by which the coated gold particles are taken into the nucleus through the pore complex.

The nuclear membrane differs in appearance from the cell membrane for the former is double (it is an "envelope") and has structures (the pore complexes) not seen in the cell membrane; and yet there are significant similarities in the transport processes across these two membranes.

In reaching this conclusion certain facts have been kept in the background. Thus the electrical properties of the nuclei in the salivary glands of *Drosophila* larvae have been referred to; but LOEWENSTEIN found the electrical properties of the frog oocyte nucleus to be different. Another example: evidence has been given for a "diffusion barrier" in the pore complex of the cell membrane; but in the pores of the nuclear membranes of the erythrocytes of frog and chicken there seems to be "continuity" between nucleoplasm and cytoplasm (DAVIES 1961). (These nuclei, it should be noted, are relatively inert metabolically.) In this preface certain facts may have been kept in the background, but no such liberties have been taken by the authors of this volume. Each presents "the whole truth and nothing but the truth"—as he sees it.

References

ALLFREY, V. G., R. MEUDT, J. W. HOPKINS, and A. E. MIRSKY, 1961: Proc. Nat. Acad. Sci. **47**, 907.

DAVIES, H. G., 1961: J. Cell Biol. **9**, 671.

IZAWA, M., V. G. ALLFREY, and A. E. MIRSKY: Unpublished experiments.

WATSON, M. L., 1959: J. Cell Biol. **6**, 154.

Electron Microscopy of the Nuclear Envelope*

By

JOSEPH G. GALL

Department of Zoology, University of Minnesota

With 15 Figures

In all organisms except the viruses, bacteria (HOPWOOD and GLAUERT 1960 a; KELLENBERGER et al. 1958), and blue green algae (HOPWOOD and GLAUERT 1960 b; RIS and SINGH 1961), the chromosomes are separated from the rest of the cell by two concentric membranes, which together constitute the nuclear *envelope*[1]. The physiological properties of the envelope, which have so far been difficult to define, clearly set a limit to the types of interactions which can occur between nucleus and cytoplasm. Except for the relatively short period during mitosis when the envelope breaks down, transport of materials to and from the chromosomes must occur across this barrier. The envelope therefore occupies a central position in all discussion of nucleo-cytoplasmic interaction. With this role in mind we may examine the structure of the nuclear envelope as revealed by the electron microscope.

Observations made with the light microscope suggested some years ago that the nucleus might be surrounded by two membranes. In particular, blisters or blebs on the surface of isolated plant nuclei were thought to result from the separation of two membranes (COHEN 1937; DANGEARD 1943). Similar blebs were also observed on animal cell nuclei under a variety of conditions (ANDERSON 1953 a; POLICARD and BESSIS 1956). Birefringence data were interpreted to indicate the presence of two membranes, one thought to be exclusively protein, the other lipoprotein (MONNÉ 1942; SCHMIDT 1939).

Evidence from electron microscopy began to accumulate in the late 1940's and early 1950's, and within a few years the main structural features

* The author's research reported in this review was supported by grants from the National Cancer Institute, U. S. Public Health Service (C-3503) and the National Science Foundation (G 10725). Present address: Department of Biology, Yale University.

[1] The term *envelope* was apparently introduced by N. G. ANDERSON. Exper. Cell Research 4. 306 (1953).

of the envelope became clear. The earliest electron microscope studies were those of CALLAN and TOMLIN (1950), who examined envelopes dissected free hand from the giant nuclei of Amphibian oocytes (*Triturus* and *Xenopus*). Their technique, which is extraordinarily valuable for studying the surface morphology of the envelope, has been used by several subsequent investigators (BOVEY 1952; GALL 1954, 1956, 1959; MERRIAM 1961 a, 1961 b, 1962).

CALLAN and TOMLIN noted two kinds of images. In one the nuclear envelope appeared to be a flat sheet pierced by circular pores about 400 Å in diameter. In the other, obtained after metal shadowing, the envelope seemed to be a continuous sheet on which were scattered raised annuli whose inner diameter corresponded to that of the pores (Fig. 4). These two images were interpreted in terms of a double membrane model, according to which the inner nuclear membrane is a continuous sheet while the outer membrane is pierced by circular pores. The annuli were thought to be drying artifacts produced from the discontinuous sheet during specimen preparation. The thickness of the envelope was estimated both by metal shadowing and by multiple beam interferometry; the values obtained were 450 Å and 500 Å respectively.

CALLAN and TOMLIN were the first to show the nuclear annuli and to suggest the existence of large pores in the envelope. Subsequent studies have required modifications in their model, however. Specifically it is now clear that the inner membrane is not a continuous sheet, nor are the annuli produced as drying artifacts.

The next important work on isolated envelopes was that of BAIRATI and LEHMANN (1952), who studied the nucleus of *Amoeba proteus*. They isolated the nucleus from fixed cells and subsequently fragmented it mechanically. The fragments were observed both before and after metal shadowing. The two sides of the fragments appeared quite different. One side (now known to be that toward the cytoplasm) was more or less continuous, but the other side was pitted with deep holes or pores. These measured about 1200 Å in diameter and were so closely spaced that the envelope displayed a most remarkable honeycomb appearance (Fig. 9, 10).

It is not surprising that BAIRATI and LEHMANN compared their porous and continuous layers with those previously described by CALLAN and TOMLIN, and that they saw in their observations further substantiation of the idea that the nuclear envelope is two layered. As luck would have it, however, BAIRATI and LEHMANN had chosen a unique envelope for their study. The porous layer which they so beautifully recorded in their micrographs is a special feature found only in *Amoeba proteus* and a few other cells (see later). The continuous layer, on the other hand, corresponds to the whole of the double nuclear envelope found in other organisms. BAIRATI and LEHMANN did not in fact see the pores and annuli which are present in their continuous layer.

BOVEY (1952) published pictures of nuclear envelopes from oocytes of the locust, *Schistocerca*, from *Amoeba discoides*, and from the gregarine, *Gregarina acridorum*, all prepared according to CALLAN and TOMLIN's procedure. His micrographs of *Schistocerca* showed evident annuli,

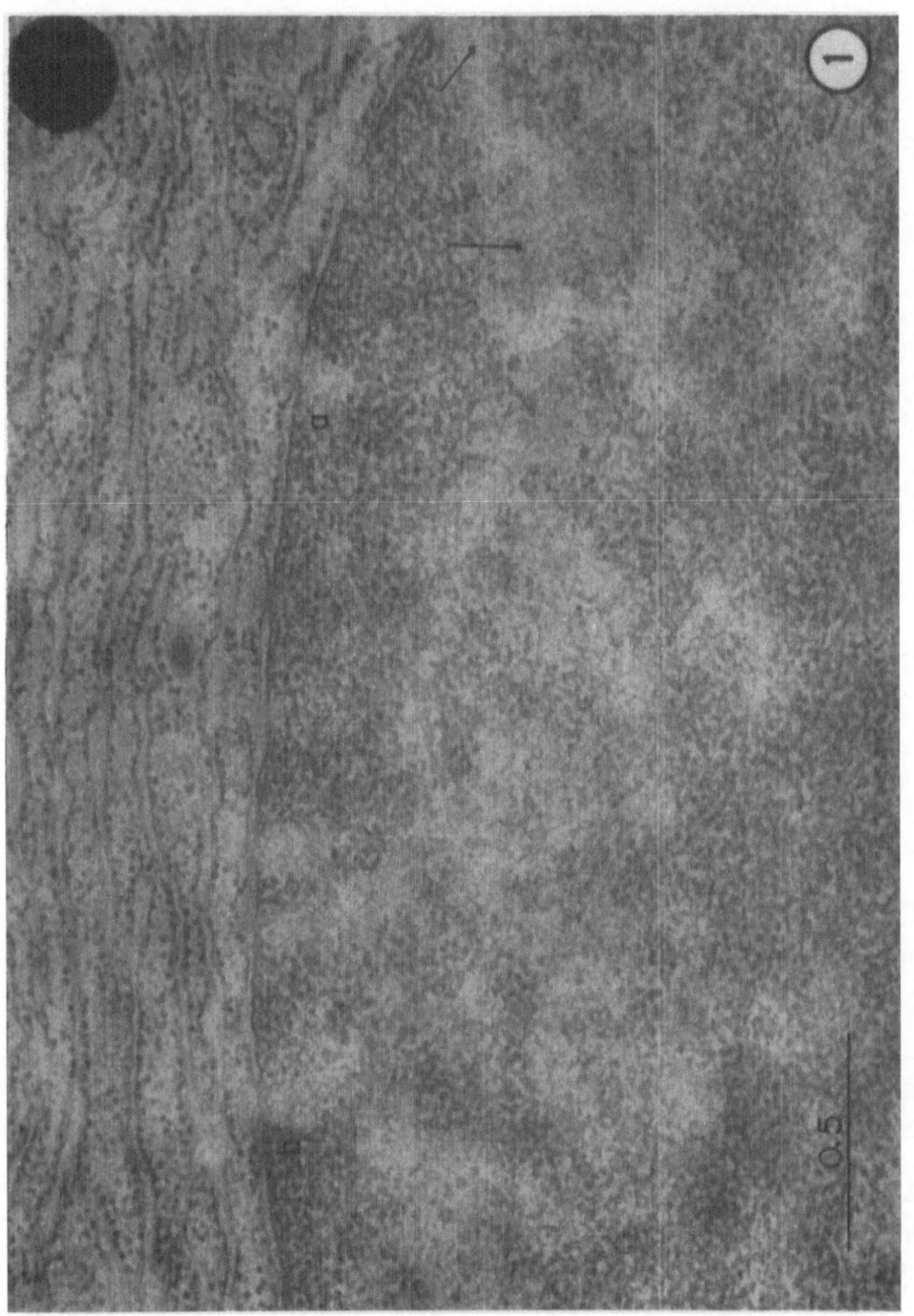

Fig. 1. An equatorial section through the nucleus of a rat pancreatic acinar cell. Sites of pores in the nuclear envelope are marked by the presence of channels of low density extending from the surface of the nucleus into its interior (arrows). "Bridges" appear across two pores (*a* and *b*) which are thought to be off-center in the section. 60,000×. From WATSON.

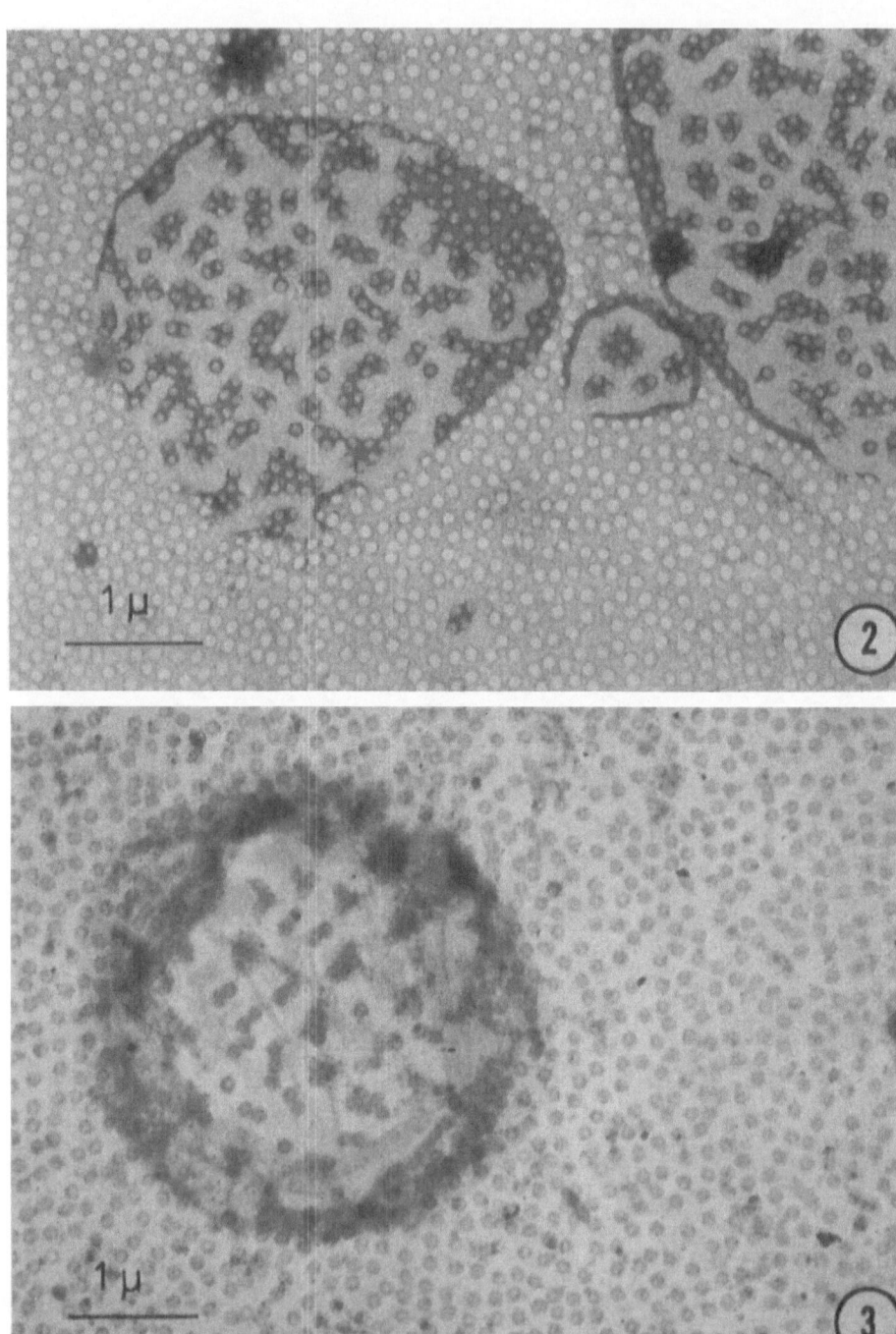

Figs. 2 and 3. Surface views of the nuclear envelope from oocytes of the newt, *Triturus*. prepared according to the technique of CALLAN and TOMLIN. The large, more or less circular areas probably represent cytoplasmic vesicles which have dried in contact with the nuclear envelope and disrupted its structure.

Fig. 2.   Fixed with 2% $KMnO_4$ unbuffered, showing prominent pores (cf. Fig. 5).   18,000×.

Fig. 3.   Fixed with 1% OsO_4 at pH 7.4, air dried, showing typical annuli.   18,000×.

suggesting that these structures might be of general occurrence. The *Amoeba* envelope appeared somewhat disorganized, but that from *Gregarina* showed obvious pores. Once again the choice of materials was unlucky, since *Gregarina* possesses an auxiliary honeycomb structure similar to that of *Amoeba proteus* (BEAMS et al. 1957); the pores demonstrated by BOVEY are very probably in the honeycomb portion!

A careful examination of these earlier papers on isolated envelopes discloses that none actually illustrate the two membranes with which we are now so familiar. This is not surprising, since the two membranes undoubtedly collapse upon one another during the drying process, and are not individually distinguishable in surface views. We must turn, then, to the earlier studies on sectioned cells to find authentic descriptions of the double nature of the envelope.

Perhaps the first convincing micrographs are those published by HART-MANN (1953) in a study of rat nerve tissue. Several of his pictures show the double envelope quite clearly, although most of the sections were not thin enough to reveal the pores and annuli. During the ensuing few years, improvements in sectioning technique came very rapidly. As a result observations on the nuclear envelope from a wide variety of organisms were published in quick succession. Important contributions were made by SJÖSTRAND and RHODIN (1953) and RHODIN (1954) working on mouse tissue, BAHR and BEERMANN (1954) studying the salivary glands of *Chironomus*, AFZELIUS (1955), who observed the envelope in sea urchin oocytes, and HAGUENAU and BERNHARD (1955), who studied several normal and cancerous mammalian tissues.

General Description of the Envelope

The two membranes which together constitute the nuclear envelope are similar in general appearance to each other and to various other membranes found in the cytoplasm (see, however, YAMAMOTO 1962). They each appear approximately 80–100 Å thick in sections of OsO_4 fixed tissue, with or without additional staining such as uranyl or lead. The separation of the two concentric membranes is somewhat variable, but is generally of the order of 100–300 Å (Figs. 1, 13, 14). The region between the membranes is referred to as the p e r i n u c l e a r s p a c e (WATSON 1955). At times each of the membranes may show evidence of a "triple layered" condition; that is, in transverse section each membrane consists of two dark lines separated by a lighter line or space. Each of these three regions is roughly 30 Å thick. Such a triple layered condition has been demonstrated for a wide variety of membranes in living systems (the so-called unit membrane of ROBERTSON (1959). The nuclear envelope, therefore, consists of two unit membranes separated by a variable, but usually narrow perinuclear space.

In transverse sections of the envelope the nuclear pores appear as discontinuities in both membranes; at the edges of the pores the inner and outer membranes are directly continuous. The diameter of the pores has been variously reported: the lowest value is about 300–400 Å (e. g. AFZELIUS

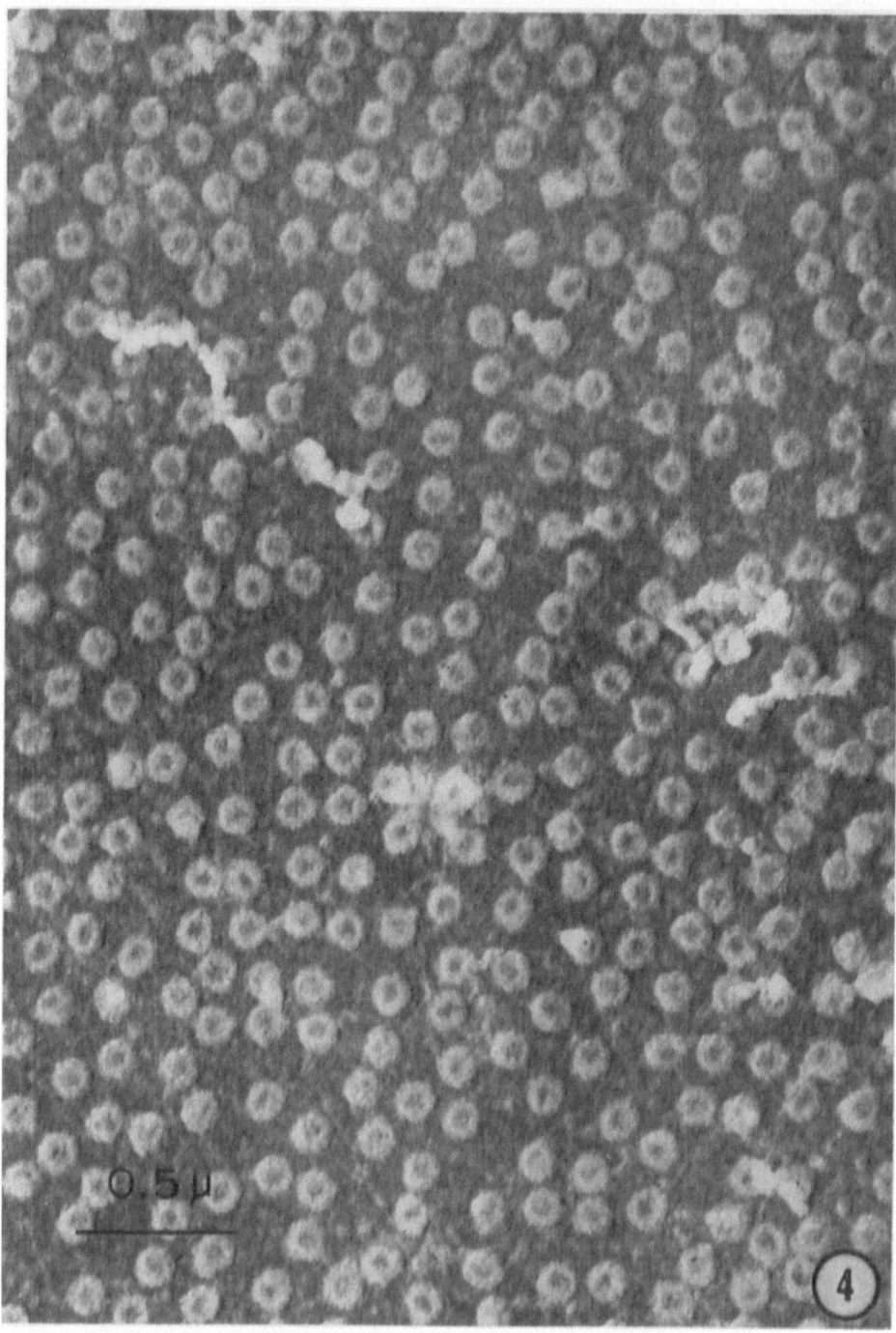

Fig. 4. Surface view of the nuclear envelope from an oocyte of the newt, *Triturus*. Spread by the technique
of CALLAN and TOMLIN, fixed with 2% OsO₄, air dried, and shadowed with chromium. 46,000×.

1955) and the highest 1000 Å (Watson 1959). Any pores which are sectioned off center will appear to have a diameter less than the true diameter. For this reason the higher figure is probably more nearly correct; it also agrees better with the value obtained from surface views of $KMnO_4$ fixed envelopes, from which the obscuring annuli are absent (Gall 1959; Luft 1956; Merriam 1961 a).

Pores have been described from many organisms, including higher plants (De 1957; Marinos 1960; Porter and Machado 1960; Whaley et al. 1960 a), Algae and Fungi (Bouck 1962; Gezelius 1959; Koehler 1962; McAlear and Edwards 1959; Mercer and Shaffer 1960; Shatkin and Tatum 1959; Turian and Kellenberger 1956) and Protozoa (Feldherr 1962; Grimstone 1959; Roth and Daniels 1962; Roth et al. 1960; Watson 1955), as well as most of the major animal phyla. They are almost certainly a universal component of the nuclear envelope and are remarkably similar in all cases that have been accurately described. The relative area occupied by the pores is difficult to estimate from transverse sections, but can be determined rather easily from surface views (Figs. 2–4). Merriam (1962) has reported that frog oocyte nuclear envelopes have between 25 and 35 pores per μ^2. Assuming a pore diameter of 1000 Å, we can calculate that roughly 25% of the nuclear surface is occupied by pores. This value is somewhat higher than the value of 5–15% estimated by Watson (1955) on the basis of 500 Å pores, and the 3% given by Barnes and Davis (1959) for acidophils.

Pores of this size, occupying such a large fraction of the nuclear surface, would provide very little barrier to free diffusion if they were open in any literal sense. Not only the largest macromolecules but even ribosomes and other small particulates could easily pass through holes 1000 Å in diameter. Most workers have felt that the envelope must do more than "keep the chromosomes in and the mitochondria out," and consequently various suggestions have been made about the patency of the pores. We shall consider here primarily evidence from electron microscopy.

It was early suggested that the pore is covered by a continuous diaphragm, somewhat thinner than the inner and outer membranes of the envelope (Afzelius 1955). Numerous published electron micrographs appear to show such a diaphragm, sometimes clearly, more often as an indistinct line across the diameter of the pore (Fig. 1). There is, however, a straightforward explanation of such "diaphragms," based on a consideration of section thickness (Barnes and Davis 1959; Watson 1959). Average sections for electron microscopy are of the order of 500 Å in thickness and even thicker ones are usable. In order that a pore appear open the section thickness must be less than the pore diameter, and furthermore, neither the front nor the back of the pore can be included in the section. Otherwise the rim of the pore will be viewed edge on and give the illusion of a diaphragm. In a careful study of this problem Watson (1959) showed that pores which appear small in transverse section (i. e., those which have been cut off center) are most often spanned by an apparent diaphragm, and their margins are apt to be indistinct. On the contrary, those pores

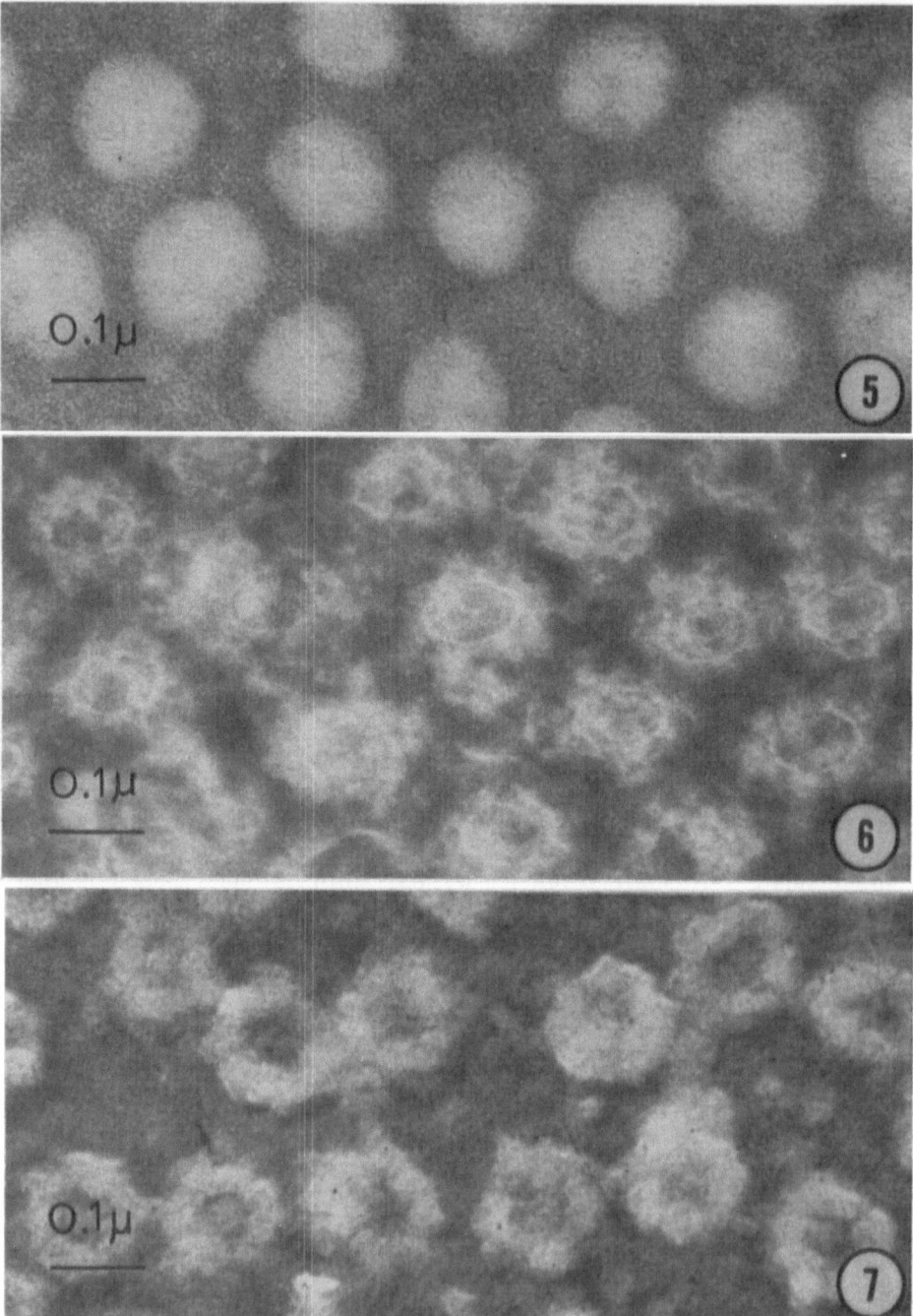

Figs. 5, 6, and 7. High resolution micrographs of small portions of the nuclear envelope from oocytes of the newt, *Triturus*. Spread by the technique of CALLAN and TOMLIN, and subsequently treated in three different ways. All reproduced at the same magnification.

Fig. 5. Fixed with 2% KMnO₄, unbuffered, and air dried. The pores are prominent but annuli are apparently absent. 140,000 ×.

Fig. 6. Fixed with 1% OsO₄ at pH 6.8, negatively stained with phosphotungstate at pH 7.0. The sharply delineated circle associated with each annulus is thought to be the rim of the pore (see text). 140,000 ×.

Fig. 7. Fixed with 2% OsO₄, air dried, shadowed with chromium. The annuli are prominent and appear to consist of irregular subunits. At least five of the annuli shown here are "plugged" with a single central granule. 140,000 ×.

which appear largest, about 1000 Å diameter, have sharp margins and no trace of a diaphragm.

An attempt was made by GALL (1959) to study the patency of the pores by a more direct method. Pieces of isolated nuclear envelope were spread over holes in a supporting carbon film, the hope being that the nuclear pores might show up as completely empty regions under these conditions. Unfortunately a thin but tough film, presumably a protein monolayer, was laid down over the entire surface of the preparation during the isolation procedure, and the experiment yielded no definite answer.

The nuclear pores show particularly well in sections of cells which have been fixed with $KMnO_4$ (LUFT 1956). Here they appear as distinct passages between the nucleus and cytoplasm, and there is no evidence of a limiting diaphragm. Isolated envelopes fixed in $KMnO_4$ show the pore regions as less dense and more distinctly bounded than the corresponding areas in osmium fixed material (Figs. 2, 5). If a limiting membrane of some sort is indeed normally present, it is not preserved by $KMnO_4$ and therefore cannot be similar in composition to the outer and inner membranes of the envelope. The absence of a limiting diaphragm after $KMnO_4$ fixation, however, could be an artifact, since other important structures (e. g. the annuli of the envelope, ribosomes, centrioles, etc.) are poorly preserved.

Although it is improbable that a distinct membrane, structurally similar to the outer and inner membranes of the envelope, covers the pores, nearly all observers have reported that the pore is "filled" with an amorphous material of greater density than the immediately adjacent nuclear or cytoplasmic areas. This material could easily contribute to the illusion of a limiting membrane, especially when viewed in thicker sections.

The Nuclear Annuli

Just as the transverse section of the nuclear envelope is characterized by two parallel lines joined in places to form the pores, so the typical surface view of the envelope displays an array of closely spaced "doughnuts" or "annuli" (Figs. 3, 4, 6, 7, 8). The annuli were first seen and named by CALLAN and TOMLIN (1950) who considered them artifacts produced when they dried their isolated material. According to their view, which on first sight seems eminently reasonable, the annulus represents the perimeter of the pore, artificially thickened by shrinkage of the outer membrane. In other words, the central clear area within the annulus would be the pore itself. Although it is true that each pore is associated with one annulus, the relationship between the two is probably more complex than CALLAN and TOMLIN envisioned.

AFZELIUS (1955) was one of the first to suggest that the annulus consists of material associated with the pore but not an integral part of the two membranes. Specifically he proposed that the "annulus" is more properly a "cylinder" extending up to 600 Å into the nucleus and 150–250 Å into the cytoplasm (Text-fig. 1 a). Most subsequent authors have likewise stressed the fact that annular material extends a considerable distance on both sides of the envelope (e.g. Swift 1958; Watson 1959; Wischnitzer 1958). However, the exact structure of the material, which often appears diffuse in transverse section, is still very much an open question.

Several authors have measured diameters of pores and annuli, and most values fall in the 500–1000 Å range. Some of the apparent variability

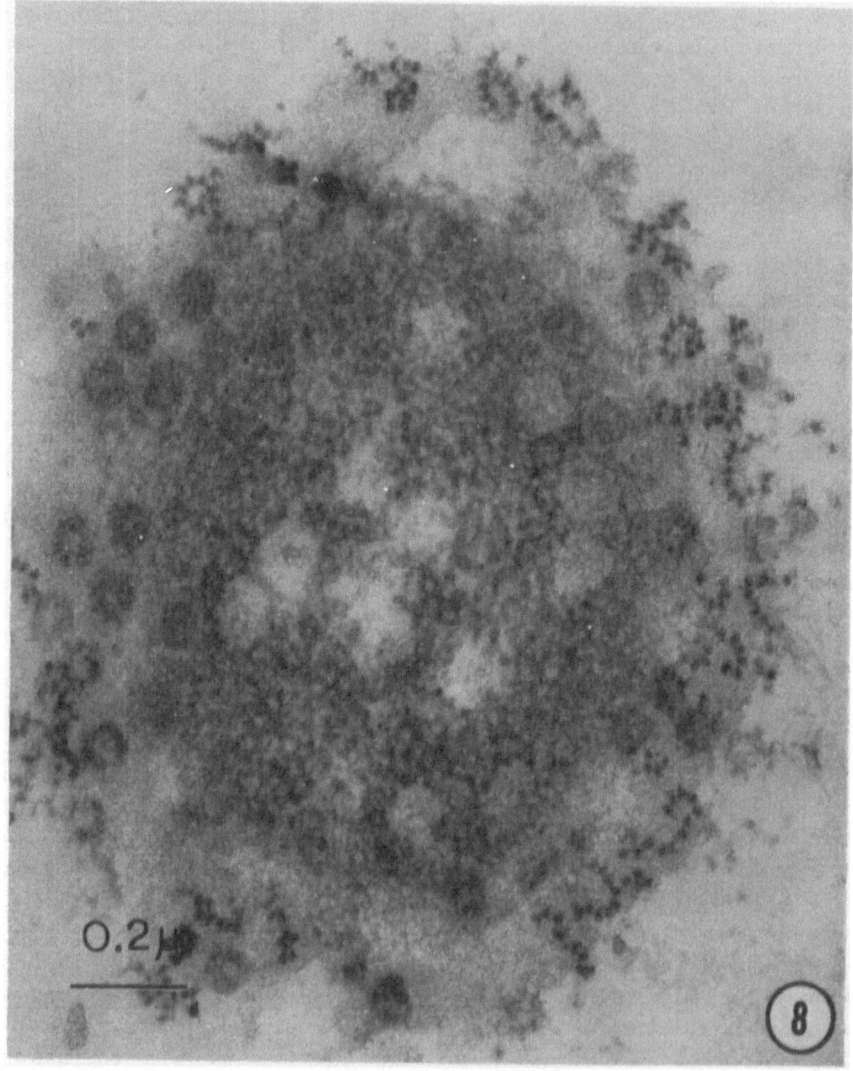

Fig. 8. Section cut tangentially to the nuclear surface, from a preparation of nuclei isolated from guinea pig liver. The outer surface of the envelope is covered with whorls and rosettes of dense particles, presumed to be ribosomes. Annuli are evident, particularly along the left hand border of the micrograph. Channels which lead from the pores deep into the nucleus are seen as clear areas near the centre (cf. Fig. 1). 80,000×. Micrograph kindly furnished by R. Maggio, P. Siekevitz, and G. Palade.

can surely be ascribed to technical factors such as method of fixation, embedding, section spreading, microscope calibration, and subjective decisions concerning the rather unsharp boundaries being measured. Whether there are, in addition, real difference from tissue to tissue or species to species is not clear. Somewhat disturbing, however, is the lack of agreement concerning the r e l a t i v e sizes of pores and annuli.

AFZELIUS made a number of careful measurements and found that the inner diameter of the annulus in starfish oocytes averaged 500–600 Å whereas the pore diameter was about 300–400Å. In his diagram he shows the cytoplasmic and nuclear portions of the annulus as equal in size but not continuous with one another (Text Fig. 1 a). WATSON (1959), on the other hand, concludes that the pores average 1000 Å in diameter and he pictures the inner diameter of the annulus as exactly coincident with the pore margin (Text-fig. 1 b). WISCHNITZER (1958 diagrams the annulus as a hollow cylinder fitting entirely within the pore; that is, the outer diameter of the annulus is made equal to the pore diameter (Text-fig. 1 c). To com-

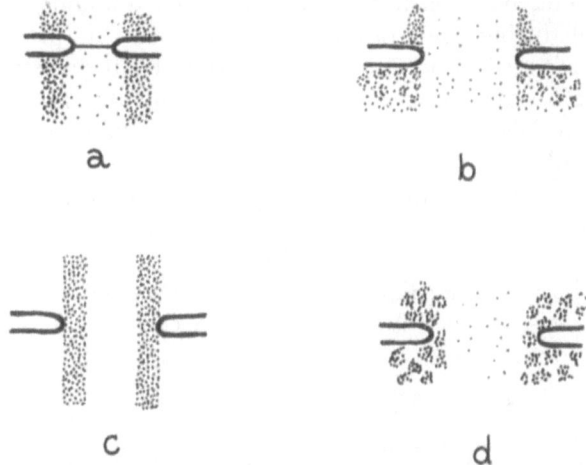

Text-fig. 1. Various interpretations of the pore-annulus relationship. Text-figs. 1 a, b. and c are modified from figures in AFZELIUS (1955), WATSON (1959), and WISCHNITZER (1958) respectively; 1 d is original. In each case the cytoplasmic surface of the nuclear envelope is toward the top of the figure. Further discussion in the text.

plete the possibilities GALL (1954, 1959) and MERRIAM (1961 a) report that the pore diameter is less than the outer, but greater than the inner diameter of the annulus (Text-fig. 1 d). These divergent opinions arise from the extreme difficulty in defining the exact boundary of the diffuse annular material, especially as seen in transverse sections of the envelope. Differences are also undoubtedly due to variations in procedure. For instance both GALL and MERRIAM base their measurements of pore diameter on surface views of $KMnO_4$ fixed envelopes, but determine the annular dimensions from OsO_4 fixed material. The annuli are not preserved by $KMnO_4$ and hence the pores are more evident after this fixative. On the other hand, we have no assurance that the pore diameter remains unaltered after $KMnO_4$.

Ideally we should like to see the annuli and pores clearly at the same time. This is not possible in transverse sections because of the indefinite outlines of the annuli. In surface views, which show the annuli adequately, the pore diameter is difficult to make out. A possible solution to this dilemma is afforded by negative staining of isolated envelopes (Fig. 6)

Here the phosphotungstate stain fills the center of the annulus and also piles up between annuli, with the result that the annulus appears as a lighter area on a dark background. An unanticipated but striking feature of negatively stained envelopes is a sharply defined circle whose diameter is greater than the inner diameter of the annulus but less than the outer diameter. This line can be interpreted as the margin of the pore proper. Contrast is probably obtained by the phosphotungstate penetrating both into the pore and also b e t w e e n the two membranes of the envelope. Under these conditions the pore margin should show up in negative contrast as a line equal to the membrane in thickness. The line in the micrograph is slightly less than 100 Å in thickness, consistent with this interpretation.

Experiments with air dried membranes, including those stained by phosphotungstate, are open to the criticism that the annular material may have collapsed during preparation. Collapsing may be prevented by the use of ANDERSON's critical point technique during the drying stage (ANDERSON 1951). Envelopes so prepared show a series of indistinct granules arranged in a circle (GALL 1956). Unfortunately the limits of the pore are not defined in such material.

In conclusion, the annuli seem to represent some type of tubular extension of non-membrane material on both the nuclear and cytoplasmic sides of the envelope. The nature of this material, whether predominately fibrous, granular, or "amorphous," is not clear. Nor is it certain just what relationship exists between the pore and its accompanying annulus.

Variations in Structure

The structure of the nuclear envelope is remarkably uniform throughout the plant and animal kingdom; only a few cell organelles, such as cilia and flagella, are so predictable in appearance. Interesting variations, however, are found in a few cell types. One of the most striking of these occurs in *Amoeba proteus*, as already briefly mentioned (Figs. 9, 10). *Amoeba* possesses a typical bilaminar nuclear envelope with pores and annuli of the usual sort (GREIDER et al. 1956; PAPPAS 1956). But in addition there is a third "layer" projecting some 2000 Å into the nucleus from the inner surface of the envelope (BAIRATI and LEHMANN 1952; GREIDER et al. 1956; HARRIS and JAMES 1952; PAPPAS 1956). This layer is constructed very much like a honeycomb, each cell of the honeycomb being associated with one nuclear pore. That is, if one imagines he is looking from the inside of the nucleus out, a pore is located at the bottom of each unit in the honeycomb. The honeycomb layer is present throughout the interphase period, but apparently disappears at the time of mitosis. The rest of the envelope, that is, the part which is identical to envelopes of other organisms, persists throughout most of mitosis (COHEN 1957; ROTH et al. 1960). The peculiar situation in *Amoeba proteus* is also found in the gregarine, *Gregarina rigida* (BEAMS et al. 1957), but is not present in the giant amoeba *Pelomyxa* (*Chaos*) (FELDHERR 1962; ROTH and DANIELS 1962), nor in the amoeboid stages of the slime molds (GEZELIUS 1959; MERCER and SHAFFER 1960). A similar honeycomb characterizes the envelope of *Endamoeba blattae,* but curiously is found on the *cytoplasmic* surface (BEAMS et al. 1959); it apparently is absent from *Entamoeba invadens* (DEUTSCH and ZAMAN 1959).

The nuclear envelope becomes modified in certain spermatids. For instance, in some organisms the envelope appears thickened in the acrosome and centriole regions (REBHUN 1957). Pores seem to be rare in late spermatids and may be absent from certain mature sperm, although this point requires further detailed study. In spermatids of the grasshopper, *Melanoplus,* portions of the chromosomal lamellae (whose structure is problematical in other respects) may be in direct continuity with a modified inner membrane (GALL and BJORK 1958).

MEEK and MOSES (1961) have described a series of modifications during the degeneration of spermatocytes in the crayfish, *Astacus.* The inner membrane of the envelope erupts into a mass of microtubules 15 to 18 mμ in diameter and up to 0.5 μ in length. At the same time the outer membrane is converted into cytoplasmic vesicles. Equally bizarre are the changes in envelope structure during normal spermiogenesis in the crayfish (MOSES 1961). Here the nuclear envelope is thrown into large folds extending deep into the cytoplasm. Eventually the cytoplasm is completely filled with complex, interdigitating extensions of the envelope. The significance of the envelope changes in normal and degenerating crayfish spermatids is not at all clear.

In Hela cells infected with adenovirus the nucleus appears to be surrounded by several concentric envelopes (GREGG and MORGAN 1959).

Breakdown and Reformation of the Envelope During Mitosis

It is well known from classical cytology that the nuclear envelope breaks down during the prophase of mitosis and is reformed in close association with the condensed chromosomes at late anaphase and telophase. Exceptions are common, such as the "intranuclear" mitosis observed in *Amoeba* and many other protozoa (e. g. CLEVELAND 1953; ROTH and DANIELS 1962; ROTH et al. 1960). The electron microscope has added interesting morphological details (BARER et al. 1959; MERRIAM 1961 b; WHALEY et al. 1960 a; MOSES 1960; PORTER and MACHADO 1960).

The breakdown seems to occur by fragmentation, followed by scattering of the pieces throughout the cytoplasm. For a while the annuli and pores can be recognized on the pieces but eventually these features are lost, or at least are difficult to recognize. The fragments then become indistinguishable from elements of the endoplasmic reticulum. Whether the total mass of the envelope fragments remains constant cannot be determined from the evidence available.

Reformation of the envelope appears in many respects to involve a reverse process. Short pieces of envelope come to lie close to the surface of the chromosomes during late anaphase and telophase. These increase progressively in number and size, presumably fusing with one another to form a continuous layer over all the chromosomes. As has been known for

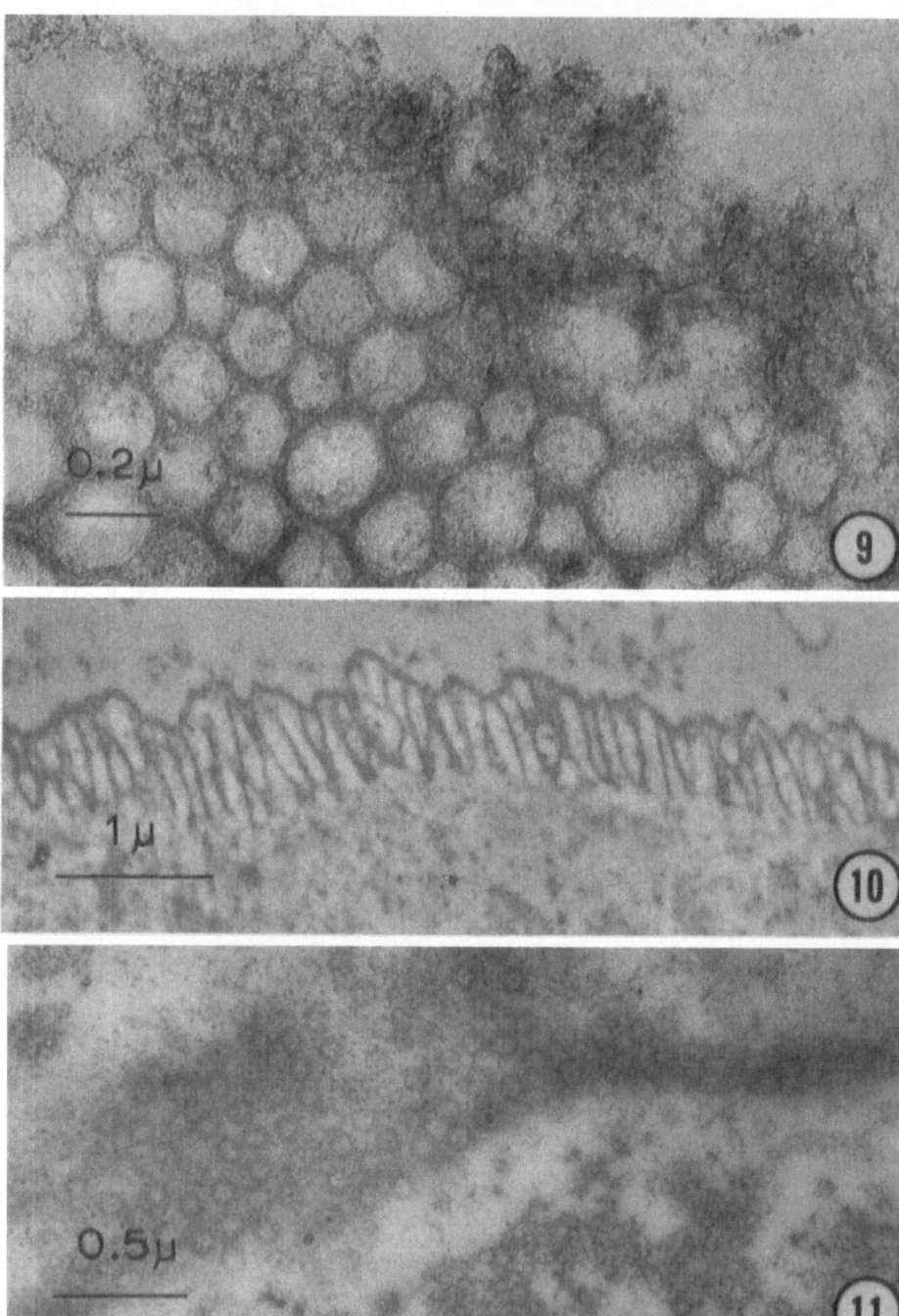

Fig. 9. The nuclear envelope of *Amoeba proteus*, from a section cut tangentially to the nuclear surface. A few annuli are visible near the top. The prominent honeycomb layer, which is attached to the inner surface of a typical bilaminar envelope, is here seen almost face-on. 68,000 ×. From ROTH, OBETZ, and DANIELS.

Fig. 10. The nuclear envelope of *Amoeba proteus*, from a section cut perpendicular to the nuclear surface. The cytoplasm is next to the top of the micrograph. The honeycomb layer extends into the nucleus from a typical bilaminar envelope, which is poorly resolved in this micrograph. 22,000 ×.

Fig. 11. Nuclear envelope from salivary gland cell of *Drosophila*. A few pores are plugged by prominent granules similar to others found free in the nucleus (lower portion of micrograph). 37.500 ×. Micrograph kindly furnished by W. BEERMANN and G. MEYER.

many years, individual chromosomes may form their own "karyomeres," complete with envelope. The karyomeres may later fuse into a single nucleus with one continuous envelope. During this process pieces of the envelope have been seen trapped within the nucleus (MOSES 1960), where they persist during the early part of the interphase.

Nucleocytoplasmic Exchange

If we are uncertain about the details of envelope morphology we know even less about the relationship between physiological behavior and envelope structure. It is clear that the properties of the envelope define the types of interactions which can occur between nucleus and cytoplasm. Yet even the most elementary questions of envelope permeability are in dispute. The existence of the pores has led naturally to the view that the nuclear envelope is a relatively open structure which should afford little resistance to the free diffusion of large or small molecules. In keeping with this viewpoint are the findings that the envelope of i s o l a t e d nuclei is permeable to large molecules such as hemoglobin and DNase (ANDERSON 1953 b; HOLTFRETER 1954). It is also well known that nuclei swell when isolated from the cytoplasm and simultaneously lose much of their protein content. Recently MACGREGOR (1962) has followed the process in isolated nuclei from newt oocytes by means of interferometry. These nuclei swell in salt solutions and lose most of their mass without obvious signs of membrane rupture. Studies with isolated nuclei however, may be a poor guide for an understanding of the normal envelope permeability. As pointed out by CHURNEY (1942), the nucleus within intact oocytes of *Arbacia* and *Nereis* responds osmotically to changes in the salt concentration of the cytoplasm. Analyses also show unequal distribution of small ions between nucleus and cytoplasm. For instance, NAORA et al. (1962) find the molar ratio of Na^+ to K^+ in frog oocyte nuclei to be 1.1, whereas it is 0.72 in the cytoplasm.

No attempt will be made here to discuss the general problem of envelope permeability. However, several observations with the electron microscope will be mentioned which bear on the question of transport of large particles across the envelope. Interactions between nucleus and cytoplasm might occur by at least three different routes: through the pores, directly across the membranes, or by way of more complex blebs and connections with cytoplasmic membranes.

Two types of nuclear envelope blebbing have been described. In the first, outpocketings of the outer membrane of the envelope occur. These vary from slight elevations of the outer membrane over a restricted area to large folds, extending deep into the cytoplasm and continuous with membranes of the endoplasmic reticulum and even with the cell surface (BENNETT 1956; GIBBS 1962; HADEK and SWIFT 1962; MCALEAR and EDWARDS 1959; MOORE and MCALEAR 1961; PALADE 1956; PARKS 1962; POLICARD and BESSIS 1956; WATSON 1955). Such outpocketings are a regular feature of certain cell types, but in other cases are either rare or entirely absent. It would be misleading, therefore, to describe the envelope as "regularly" in continuity

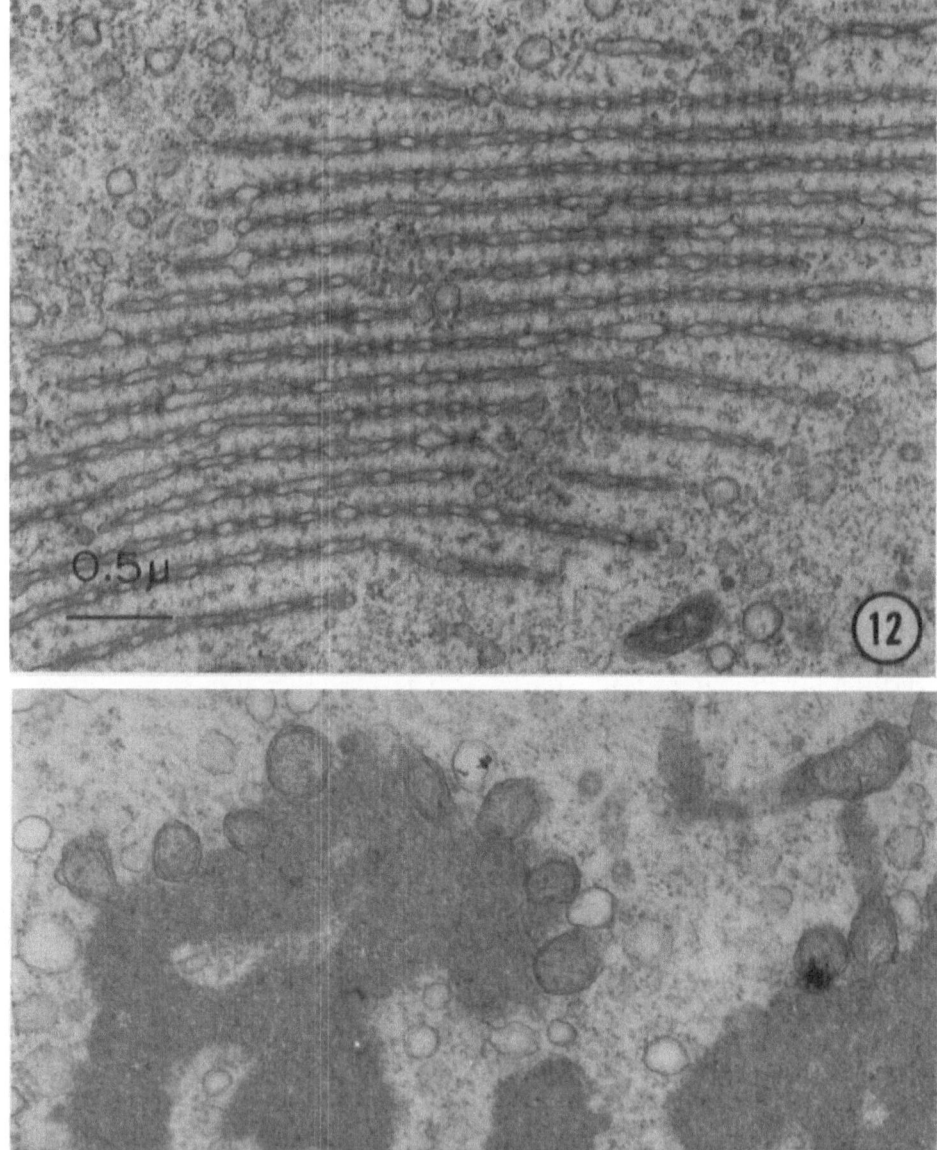

Fig. 12. A stack of annulate lamellae in the oocyte cytoplasm of a tadpole, *Rana clamitans*. Note the dense material which fills the pores and extends as indistinct projections between successive layers. These projections probably represent sections through annuli (cf. Text-Figure 1).   30,000 ×.   Micrograph kindly furnished by O. L. Miller.

Fig. 13. Masses of dense material in the oocyte cytoplasm of a tadpole, *Rana clamitans*. Note the fingerlike projections which connect these masses to pores in the nuclear envelope, suggesting that material may be secreted through the pores into the cytoplasm.   30,000 ×,   Micrograph kindly furnished by O. L. Miller.

with the membrane systems of the cytoplasm. The existence of such nuclear blebs has led to the suggestions that cytoplasmic membranes may derive ultimately from the nucleus, and indeed may be endowed with highly specific properties as a result of their previous association with the chromosomes (SWIFT 1958). Equally cogent arguments can be advanced in favor of the view that cytoplasmic membranes are derived from the cell surface by pinching off of pinocytotic vesicles (BENNETT 1956; HOLTER 1961). Since the cytoplasmic membranes usually lack pores and annuli, it is not easy to imagine their coming directly from the nuclear envelope. We can, of course, assume a change in structure as membranes are given off by the nucleus, but at present we lack any solid evidence.

The second type of blebbing involves outpocketing of both membranes of the envelope (CLARK 1960). Such blebs have been described by GAY (1955, 1956 b) and KAUFMANN and GAY (1958) in a careful study involving the use of serial sections. The blebs were found to be common on nuclei of third instar *Drosophila* larvae, but rarer or absent at other stages. The outpocketings are often demonstrably associated with a chromosome band. These authors believe that blebs are pinched off from the nucleus, carrying with them specific products from the chromosomes. Such a mechanism for transport of nuclear material to the cytoplasm does not appear to be a universal feature of cells, although it may be important in certain cases.

Many authors have speculated that high molecular weight substances, or even organized streams of material pass through the nuclear pores. WATSON (1955, 1959) has shown in several cells of the rat that distinct channels lead from the pores deep into the nucleus (Figs. 1, 8). The material within the channel is less dense than the surrounding chromosomal material and appears to consist of very delicate chains or filaments. Such channel material extends through the pores into the cytoplasm a short distance where it can be distinguished by its texture and the absence of cytoplasmic particulates such as ribosomes. Various other authors have noted the presence of filamentous material projecting into the cytoplasm from the pores (e. g. SWIFT 1958).

The most striking evidence for extrusion of nuclear material through the pores comes from observations on several types of oocytes. POLLISTER, GETTNER, and WARD (1954) described dense cytoplasmic material in association with the pores in developing frog oocytes, and suggested that material was passing from nucleus to cytoplasm. MILLER (1962) has recently studied the same cells. At various points in the cytoplasm around the periphery of the nucleus are found clumps of dense material, from which quite uniform fingers extend toward the nuclear pores. The diameter of the fingers is equal to the pore diameter and it appears that many if not all of the projections end exactly at a pore (Fig. 13). The dense clumps are readily visible in the light microscope and have been shown to contain RNA. Mitochondria are found in large numbers in these regions, embedded in the basophilic masses. Very similar observations have been made on oocytes of the bug *Rhodnius* by ANDERSON and BEAMS (1956) and on spider oocytes by SWIFT (1958) and ANDRÉ and ROUILLER (1957). In none of these cases is it certain which way the material is moving, although the usual interpretation involves extrusion of nuclear RNA. Recent experiments of FELDHERR (1962; FELDHERR and MARSHALL 1962) provide direct evidence for the passage of particulate matter, in this case from cytoplasm to nucleus. FELDHERR injected colloidal gold particles, 25–55 Å in diameter, into living *Amoebae*, which were then fixed and sectioned. After 24 hours particles were found in both nucleus and cytoplasm, surprisingly at a higher concentration in the nucleus. Even after 10 minutes there were many particles in the nucleus. At one and two mintues after injection most of the gold was still in the cytoplasm, although individual particles were frequently found within the annuli of the envelope. In many cases single gold particles were found in the exact center of the annulus, as if passage through the pore is limited to a small central channel.

Centrally located granules have been found normally within annuli of a wide variety of cells (e. g. AFZELIUS 1955; GALL 1954; MERRIAM 1962; SWIFT 1958; WISCHNITZER 1958). These have been interpreted either as part of the pore complex or as material passing through the pore. In light of FELDHERR's study, we may reasonably suppose that such particles are in actual transit (Fig. 7, 11).

In summary, there is both circumstantial and direct evidence that material can pass through the pores of the nuclear envelope. There is also evidence of an association between the nuclear envelope and membrane systems of the cytoplasm, and it is possible that transport of materials may involve the membranes. In addition small molecules can probably penetrate the membranes directly, although there is, of course, no evidence from electron microscopy to support this latter mechanism. The apparent permeability properties of the nuclear envelope will very according to the nature and size of the particles considered, which of several routes is actually used in passage through the envelope, and the time course of the experiment. Molecules which can pass through the envelope of an isolated nucleus (whose pores may be artificially open) may not penetrate so readily when the pores are filled with channel material; although, given time, they might become incorporated into a slowly moving stream and pass through.

The Annulate Lamellae

The nuclear envelope, with its conspicuous pores and annuli, is usually distinguishable from such cytoplasmic membranes as the endoplasmic reticulum, the Golgi material, pinocytotic vesicles, etc. However, certain cell types regularly possess cytoplasmic membranes identical to the nuclear membranes, but not in direct continuity with them. These membranes have been called periodic lamellae (REBHUN 1956) or annulate lamellae (SWIFT (1956). They exist as double layered sacs of the same dimensions as the nuclear envelope and possessing typical pores and annuli. They may be found singly or in stacks of up to a dozen or more double layered units (Figs. 12, 14). When found in stacks the component lamellae show extraordinary alignment, the pores and annuli of successive layers being in

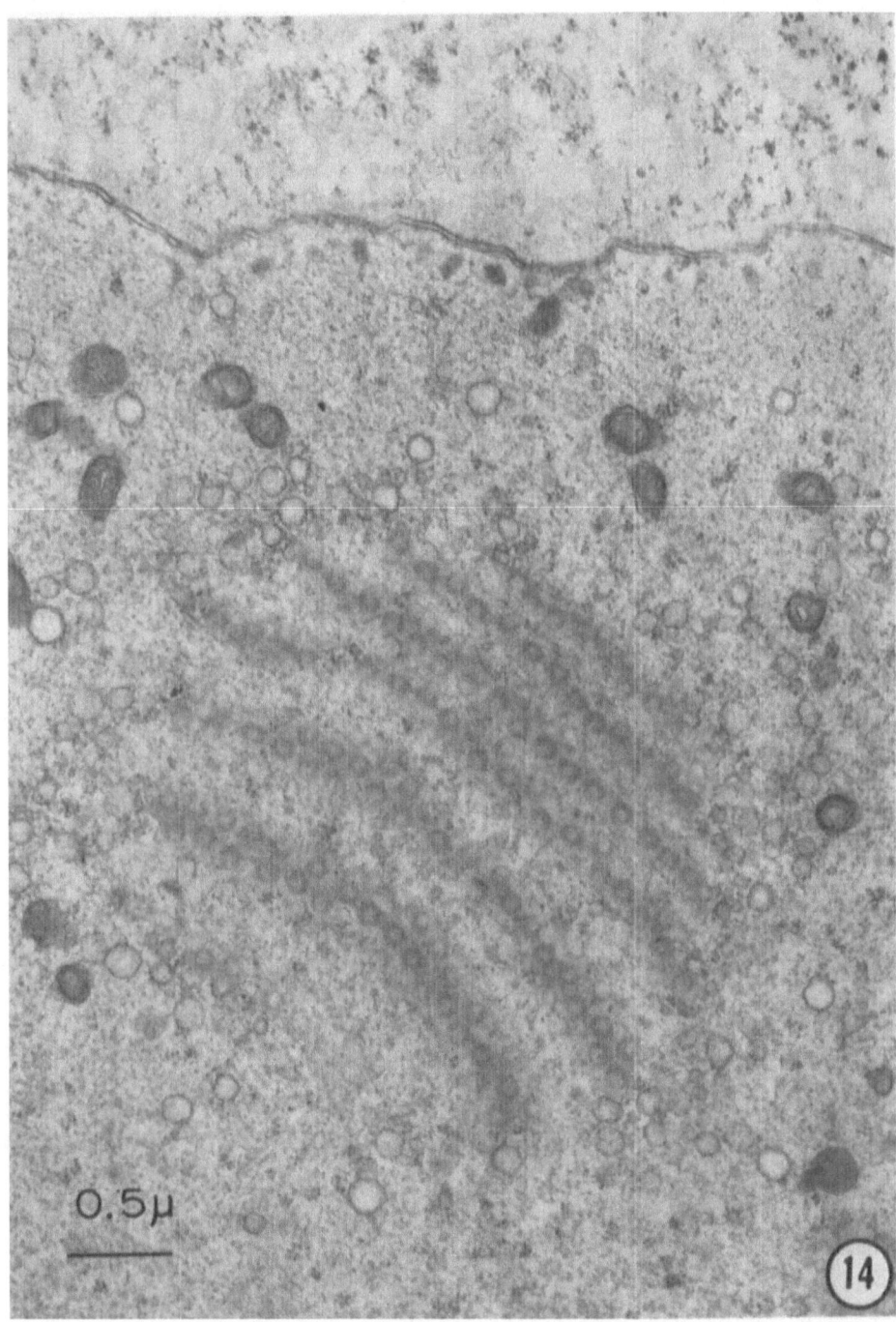

Fig. 14. Annulate lamellae in the oocyte cytoplasm of a tadpole, *Rana clamitans*. Here the successive layers are sectioned tangentially to their surfaces, and the annuli are distinct. The nuclear envelope extends horizontally across the top of the micrograph. 30,000×. Micrograph kindly furnished by O. L. MILLER.

close registry with one another. In transverse section denser material (part of the annuli?) seems to extend between pores in successive layers, giving the whole structure a regular periodic appearance (Fig. 12). The annulate lamellae are occasionally oriented parallel to the nucleus, so that the nuclear envelope forms the innermost layer in a stack. At other times the stacks show little or no obvious association with the nucleus. Annulate lamellae have been described primarily from germ cells, both spermatocytes (Ruthmann 1958; Swift 1956) and oocytes (Afzelius 1957; Merriam 1959; Rebhun 1956, 1961; Wischnitzer 1960), although occasionally they have been seen elsewhere (Swift 1956). Stacks of annulate lamellae are frequently large enough to be seen as dense "granules" in sections observed by light microscopy. Cytochemical studies have shown that such stacks are strongly basophilic, even though they may be relatively free of ribosome-like particles (Rebhun 1961; Ruthmann 1958). This observation suggests that RNA may exist in the membranes, or in the annuli, in a form different from the typical 150 Å particles. Annulate lamellae may be in direct continuity with typical non-annulated membranes of the endoplasmic reticulum. Ruthmann (1958) has described such connections as well as associations with even more complex membrane systems.

Several workers have suggested that the nuclear envelope serves as a template for the formation of annulate lamellae, or that the lamellae arise by simple delamination from the nuclear surface. The best documented evidence, however, suggests a somewhat different mechanism. In oocytes of the salamander, *Necturus*, Kessel (1963) has shown that numerous cytoplasmic vesicles are given off as small blebs from the outer nuclear membrane. These travel peripherally in the cytoplasm where they align and eventually coalesce to form the layers of annulate lamellae. Just how the pores and annuli arise during this process is not clear, since the individual vesicles show no signs of them. The function of the annulate lamellae is unknown.

References

Afzelius, B., 1955: Exper. Cell Res. 8, 147.
— 1957: Z. Zellforsch. 45, 660.
Anderson, E., and H. W. Beams, 1956: J. Biophys. Biochem. Cytol. 2 (suppl.), 439.
Anderson, N. G., 1953 a: Exper. Cell Res. 4, 306.
— 1953 b: Exper. Cell Res. 5, 361.
Anderson, T. F., 1951: Trans. N. Y. Acad. Sci. 13, 130.
André, J., and C. Rouiller, 1957: Proc. Stockholm Conf. Electron Microscopy, 1956, p. 162. Almqvist and Wiksell, Stockholm, and Academic Press, New York.
Bahr, G., and W. Beermann, 1954: Exper. Cell Res. 6, 519.
Bairati, A., and F. E. Lehmann, 1952: Experientia 8, 60.
Barer, R., S. Joseph, and G. A. Meek, 1959: Exper. Cell Res. 18, 179.
Barnes, B., and J. M. Davies, 1959: Ultrastruct. Res. 3, 131.
Beams, H., T. Tahmisian, R. Devine, and E. Anderson, 1957: Exper. Cell Res. 13, 200.
Beams, H. W., T. Tahmisian, R. Devine, and E. Anderson, 1959: Exper. Cell Res. 18, 366.
Bennett, H. S., 1956: J. Biophys. Biochem. Cytol. 2 (suppl.), 99.
Bouck, G. B., 1962: J. Cell Biology 12, 553.
Bovey, R., 1952: J. Roy. Microscop. Soc. 72, 56.
Brandt, P. M., and G. D. Pappas, 1959: J. Biophys. Biochem. Cytol. 6, 91.
Callan, H. G., and S. G. Tomlin, 1950: Proc. Roy. Soc. London B 137, 367.

CHURNEY, L., 1942: Biol. Bull. 82, 52.
CLARK, W. H., Jr., 1960: J. Biophys. Biochem. Cytol. 7, 345.
CLEVELAND, L. R., 1953: Trans. Amer. Phil. Soc. 43, 809.
COHEN, A. I., 1957: J. Biophys. Biochem. Cytol. 3, 859.
COHEN, I., 1937: Protoplasma 27, 484.
DANGEARD, P., 1943: Compt. Rend. Soc. Biol. 137, 233.
DE, D. N., 1957: Exper. Cell Res. 12, 181.
DEUTSCH, K., and V. ZAMAN, 1959: Exper. Cell Res. 17, 310.
FELDHERR, C. M., 1962: J. Cell Biology 14, 65.
— and A. B. FELDHERR, 1960: Nature 135, 250.
— and J. M. MARSHALL, 1962: J. Cell Biology 12, 640.
GALL, J. G., 1954: Exper. Cell Res. 7, 197.
— 1956: J. Biophys. Biochem. Cytol. 2 (suppl.), 393.
— 1959: J. Biophys. Biochem. Cytol. 6, 115.
— and L. BJORK, 1958: J. Biophys. Biochem. Cytol. 4, 479.
GAY, H., 1955: Proc. Nat. Acad. Sci. 41, 370.
— 1956 a: J. Biophys. Biochem. Cytol. 2 (suppl.), 407.
— 1956 b: Cold Spring Harbor Symposia 21, 257.
GEZELIUS, K., 1959: Exper. Cell Res. 18, 425.
GIBBS, S. P., 1962: J. Cell Biology 14, 433.
GREGG, M. B., and C. MORGAN, 1959: J. Biophys. Biochem. Cytol. 6, 539.
GREIDER, M. H., W. J. KOSTIR, and W. J. FRAJOLA, 1956: J. Biophys. Biochem.
    Cytol. 2 (suppl.), 445.
GRIMSTONE, A. V., 1959: J. Biophys. Biochem. Cytol. 6, 569.
HADEK, R., and H. SWIFT, 1962: J. Cell Biology 13, 445.
HAGUENAU, F., and W. BERNHARD, 1955: Bull. Cancer 42, 537.
HARRIS, P., and T. JAMES, 1952: Experientia 8, 384.
HARTMANN, J. F., 1953: J. Comp. Neurology 99, 201.
HOLTER, H., 1961: Biological Structure and Function. p. 157 (ed. by T. W. GOODWIN
    and O. LINDBERG). Academic Press.
HOLTFRETER, J., 1954: Exper. Cell Res. 7, 95.
HOPWOOD, D. A., and A. M. GLAUERT, 1960 a: J. Biophys. Biochem. Cytol. 8, 267.
— — 1960 b: J. Biophys. Biochem. Cytol. 8, 813.
KAUFMANN, B. P., and H. GAY, 1958: The Nucleus 1, 57.
KELLENBERGER, E., A. RYTER, and J. SECHAUD, 1958: J. Biophys. Biochem. Cytol. 4, 671.
KESSEL, R. G., 1963: J. Cell Biol. 19, 391.
KOEHLER, J. K., 1962: J. Ultrastruct. Res. 6, 432.
LUFT, J., 1956: J. Biophys. Biochem. Cytol. 2, 799.
MACGREGOR, H. C., 1962: Exper. Cell Res. 26, 520.
MARINOS, N. G., 1960: J. Ultrastruct. Res. 3, 328.
McALEAR, J. H., and G. A. EDWARDS, 1959: Exper. Cell Res. 16, 689.
MEEK, G., and M. MOSES, 1961: J. Biophys. Biochem. Cytol. 10, 121.
MERCER, E. H., and B. M. SHAFFER, 1960: J. Biophys. Biochem. Cytol. 7, 553.
MERRIAM, R., 1959: J. Biophys. Biochem. Cytol. 5, 117.
— 1961 a: J. Biophys. Biochem. Cytol. 11, 559.
— 1961 b: Exper. Cell Res. 22, 93.
— 1962: J. Cell Biol. 12, 79.
MILLER, O. L., 1962: Electron Microscopy. Proc. 5th Intern. Congress Electron
    Microscopy 2, NN 8.
MOLLENHAUER, H. H., 1959: J. Biophys. Biochem. Cytol. 6, 431.
MONNÉ, L., 1942: Arkiv Zool. 34 B, No. 2.
MOORE, R. T., and J. H. McALEAR, 1961: Exper. Cell Res. 24, 588.
MOSES, M., 1960: 4th Intern. Congr. Electron Microscopy, II, p. 230. Springer, Berlin.
— 1961: J. Biophys. Biochem. Cytol. 10, 301.
NAORA, H., H. NAORA, M. IZAWA, V. ALLFREY, and A. MIRSKY, 1962: Proc. Nat. Acad.
    Sci. 48, 853.
PALADE, G. E., 1956: J. Biophys. Biochem. Cytol. 2 (suppl.), 85.
PALAY, S. L., 1960: J. Biophys. Biochem. Cytol. 7, 391.
PAPPAS, G. D., 1956: J. Biophys. Biochem. Cytol. 2 (suppl.), 431.
PARKS, H. F., 1962: J. Cell Biol. 14, 221.
POLICARD, A., and M. BESSIS, 1956: Exper. Cell Res. 11, 490.
POLLISTER, A., M. GETTNER, and R. WARD, 1954: Science 120, 789.
PORTER, K. R., and R. D. MACHADO, 1960: J. Biophys. Biochem. Cytol. 7, 167.

REBHUN, L. I., 1956: J. Biophys. Biochem. Cytol. 2, 93.
— 1957: J. Biophys. Biochem. Cytol. 3, 509.
— 1961: J. Ultrastruct. Res. 5, 208.
RHODIN, J., 1954: Correlation of ultrastructural organization and function in normal and experimentally-changed proximal convoluted tubule cells of the mouse kidney. Stockholm.
RIS, H., and R. N. SINGH, 1961: J. Biophys. Biochem. Cytol. 9, 63.
ROBERTSON, J. D., 1959: Biochem. Soc. Symp. 16, 3.
ROTH, L. E., and E. W. DANIELS, 1962: J. Cell Biology 12, 57.
— S. W. OBETZ, and E. W. DANIELS, 1960: J. Biophys. Biochem. Cytol. 8, 207.
RUTHMANN, A., 1958: J. Biophys. Biochem. Cytol. 4, 267.
SCHMIDT, W. J., 1959: Protoplasma 32, 193.
SHATKIN, A. J., and E. L. TATUM, 1959: J. Biophys. Biochem. Cytol. 6, 423.
SJÖSTRAND, F., and J. RHODIN. 1953: Exper. Cell Res. 4, 426.
SWIFT, H., 1956: J. Biophys. Biochem. Cytol. 2 (suppl.), 415.
— 1958: in: A Symposium on the Chemical Basis of Development. (Ed. W. D. McElroy and B. Glass.) The Johns Hopkins Press, p. 174.
TURIAN, G., and E. KELLENBERGER: Exper. Cell Res. 11, 417.
WATSON, M., 1955: J. Biophys. Biochem. Cytol. 1, 257.
— 1959: J. Biophys. Biochem. Cytol. 6, 147.
WHALEY, W., H. MOLLENHAUER, and J. LEECH, 1960 a: Amer. J. Botany 47, 401.
— — — 1960 b: J. Biophys. Biochem. Cytol. 8, 233.
WISCHNITZER. S., 1958: J. Ultrastruct. Res. 1, 201.
— 1960: J. Biophys. Biochem. Cytol. 8, 558.
YAMAMOTO, T., 1962: Electron Microscopy. Proc. 5th Intern. Congr. Electron Microscopy 2, LL 6.

Permeability of the Nuclear Membrane as Determined with Electrical Methods

By

WERNER R. LOEWENSTEIN

Department of Physiology, Columbia University, College of Physicians and
Surgeons, New York

With 9 Figures

Our knowledge of the permeability and electrochemical properties of
cell membranes has been greatly advanced in the past 15 years by the
development of direct intracellular electric measuring techniques. It is
possible, for example, to drive a minute, but accurately measured ion
current across a membrane, and to measure the membrane conductance
or "permeability" to this current. Such methods used until quite recently
only in studies of surface membranes, have now been applied to nuclear
membranes. This article summarizes some of the results.

The main features of the approach appear in Fig. 1. Two micropipettes
of tip diameter below 0.5 μ are inserted into the nucleus. One serves to pass
current through the nuclear membrane, and the other to measure the
resulting voltage drop. The nuclear and cell membranes seal well around
such pipettes; there is no sign of current leakage. There are also no visible
structural changes, such as have been observed upon puncturing nuclear
membranes with instruments of larger tips (CHAMBERS and FELL 1931; KOPAC
and MATEYKO 1958). The density of current is rather uniform over the
nuclear membrane, the membrane area is readily measured, and the
transverse resistance of unit membrane area can be determined with a
high degree of accuracy. Measurements are done *in situ* with pulses of
current as small as 10^{-8} amp. and as short as 5 msec. Changes in normal
ion composition around the membrane due to transfer of ionic charge are
thus extremely small. This minimizes the risk of causing changes in membrane properties and offers a significant advantage over other methods in
which, as discussed in a subsequent article of this series (p. 35), the quantity
or nature of the probing agent itself may alter the permeability of the
membrane. A shortcoming of the method is that it yields values of lumped
permeability only; in its present form it provides no information about

the identity of the ions which carry the current. (For a detailed description of the technique see Loewenstein and Kanno, 1963 b.)

The internal recording technique as applied to the nucleus is still in its early stages and cannot yet be applied but to large nuclei. The results obtained in large interphase nuclei of salivary gland cells of Drosophila and Chironomid larvae will here be briefly reviewed.

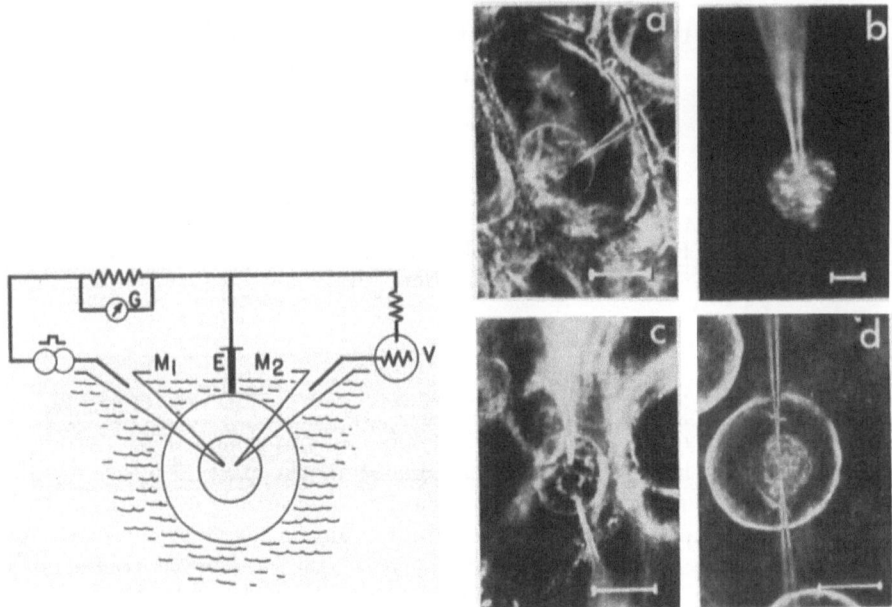

Fig. 1. Left. Diagram of set-up for potential and conductance measurements. Square pulses of current are passed through the cell and nucleus membranes between an intranuclear electrode M_1 (an electrolyte-filled micropipette) and an electrode E located in the fluid outside the cell; the resulting voltage drops and resting potential are recorded between another intranuclear electrode M_2 and E.

Right. Dark field photomicrographs of unstained nuclei impaled with microelectrodes. a, c, nuclei of *Drosophila flavorepleta* salivary gland cells *in situ*. b, semi-isolated; d, of oocytes of *Xenopus laevis*. Calibration, a, b, c: 25 μ. d; 100 μ. (From Loewenstein and Kanno, 1963a.) Reproduced by permission of the Journal of Cell Biology.

Gland Cell Nuclei

Membrane potential. When the nucleus and cytoplasm of a Drosophila salivary gland cell are explored with potential recording micropipettes, the existence of a potential difference between the two phases is immediately noticed (Fig. 2). The potential arises clearly at the phase boundary; there are no detectable potential differences within the nucleoplasm or cytoplasm. The potential is associated with the nuclear surface, presumably with the surface complex which under the electron microscope appears as a double-layered membrane. This is brought out with particular clarity in experiments in which the nuclear membrane is ruptured experimentally, while the potential is continuously monitored. The potential is then found to decline to zero (Fig. 3). (Loewenstein and Kanno 1962, 1963 a, 1963 b.)

Like the cell membrane, the nuclear membrane appears thus to act as a charge separator, the inside of the membrane being negative relative to the outside. The average potential difference across the nuclear membrane is 15 mV. This implies that the nuclear membrane must normally sustain very high electrical gradients (on the order of 10.000 V per cm.) and indicates that it is a good insulator.

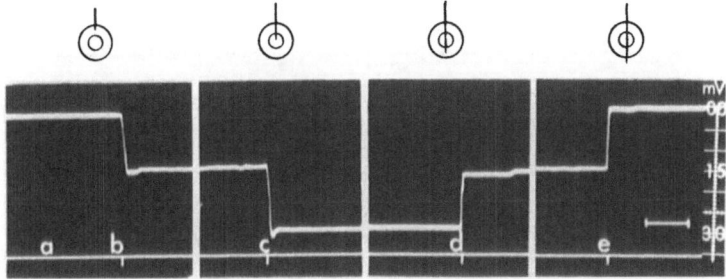

Fig. 2. Potentials across cell and nucleus membrane (*Drosophila flavorepleta*). A potential-recording microelectrode is advanced progressively in the direction cell exterior-nucleus. Upper oscilloscope beam records the potential (downward negative) as the electrode tip is *a*. outside the cell; *b*. entering the cell membrane; *c*, entering the nucleus; *d*, leaving the nucleus; *e*, emerging into the cell exterior. Reference electrode is in the cell exterior. Time calibration 0.05 sec.; film interrupted for about 1 sec. in between photographs. (From LOEWENSTEIN and KANNO, 1963b.) Reprinted by permission of the Journal of General Physiology.

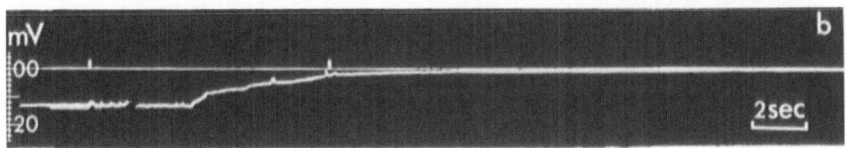

Fig. 3. The nucleus potential after destruction of the nuclear membrane (*Drosophila flavorepleta*). A microelectrode inside the nucleus records continuously the nuclear membrane potential while a hole is drilled into the nuclear membrane. Beginning and end of drilling period are marked on upper beam. The recording electrode stays inside the nucleus throughout the experiments. (From LOEWENSTEIN and KANNO, 1963b.) Reprinted by permission of the Journal of General Physiology.

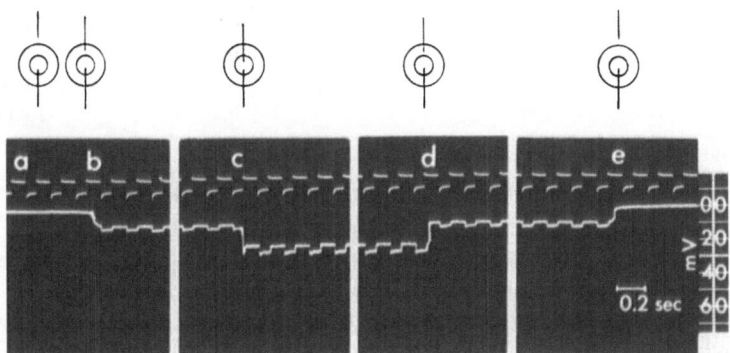

Fig. 4. Nucleus and cell membrane resistance (*Drosophila flavorepleta*). Current pulses of constant strength (upper beam) are passed through the nucleus and cell membranes between a microelectrode placed inside the nucleus and an electrode in the extracellular fluid. The membrane potential (lower beam) is recorded continuously between a reference electrode in the extracellular fluid and a second microelectrode, as pictured in the diagram of Fig. 1. moving in the direction extracellular fluid-nucleus and back. Recording electrode *a*, *e*, in extracellular fluid; *b*, *d*, entering cytoplasm: *c*, entering nucleoplasm. (From LOEWENSTEIN and KANNO, 1963b.) Reprinted by permission of the Journal of General Physiology.

Membrane Resistance. The nuclear membrane of *Drosophila* gland cell nuclei behaves, indeed, like a dielectric. It has a high electrical resistance and a high capacitance (Table 1; Figs. 4 and 5). Both are clearly associated with the nuclear membrane, for if this membrane is ruptured, the capacitance becomes immeasurably small and the resistance drops to a fraction of its original value which corresponds approximately to the resistance between the micropipettes in this system in absence of a membrane (Fig. 5 b).

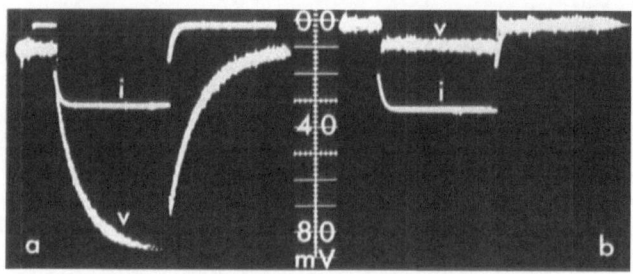

Fig. 5. Nuclear membrane resistance. Square pulses of 1.5×10^{-6} A (i) are passed across the membrane of a semi-isolated nucleus (*Drosophila flavorepleta*) (current density 39.5×10^{-3} A/cm²), the resulting voltage drop (*v*) is recorded before (*a*) and after (*b*) drilling a hole into the nuclear membrane. Note the large capacitative component of membrane impedance in the intact membrane. The capacitance of this nuclear membrane amounts to 1.5×10^{-8} F and the resistance to 48,000 Ω. Time calibration: 1 msec. (From LOEWENSTEIN and KANNO, 1963 b.) Reprinted by permission of the Journal of General Physiology.

Table 1. *Potential and resistance of nuclear membranes.*

Cell type	Nuclear Membrane potential mv	Nuclear Membrane resistance Ω cm²
Gland cell *(Drosophila flavorepleta)*	14.5 ± 1.1	1.5 ± 0.4
Gland cell *(Chironomus thummi)*	≈ 15	≈ 2
Oocyte[1] *(Xenopus laevis)*	0 ± 0.1	< 0.001
Oocyte *(Triturus viridescens)*	0 ± 0.3	< 0.001
Oocyte *(Asterias forbesi)*	0 ± 0.3	< 0.001
Oocyte *(Nereis limbata)*	0 ± 0.3	< 0.001
Oocyte *(Spisula solidissima)*	0 ± 0.3	< 0.001
Oocyte *(Hydractinia echinata)*	0 ± 0.3	< 0.001

Mean values with their standard errors.

[1] Transparent oocytes of diameter between 80 and 300 μ only.

Measures of nuclear membrane resistance have been obtained from current voltage relations in which the current was varied over a wide range in both directions through the membrane. In contrast to the surface membrane of many excitable cells, the nuclear membrane shows little or no rectification and gives no sign of excitation over the entire range of current that can be used without damage to the membrane (Fig. 6). Recent studies on salivary gland cell nuclei of Chironomid larvae (late prepupae) showed similar membrane potentials (H. KROEGER. personal communication) and membrane resistances (HIGASHINO and LOEWENSTEIN, unpublished data).

To sum up. the nuclear membrane of these gland cell nuclei may be pictured as a spherical surface of low ionic conductance. which separates two aqueous phases of high ionic conductance and unequal charge. The potential difference across the membrane is about 15 mV and the resistance of the order of 1 Ω cm.², while there is no detectable potential difference and comparatively negligible resistance within either phase.

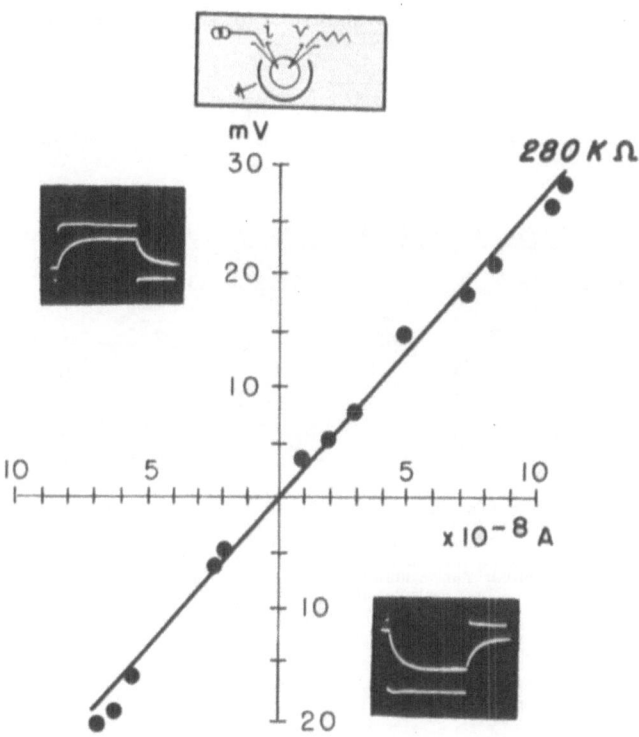

Fig. 6. Current voltage relation in a nuclear membrane (semi-isolated nucleus. *Drosophila flavorepleta*). Abscissae: total nuclear membrane current; inward current, left. Ordinates: "steady-state" potential, hyperpolarization downwards. Nuclear membrane area 1.6 × 10⁻⁵ cm. (From LOEWENSTEIN and KANNO, 1962.) Reprinted by permission of Nature.

Oocyte Nuclei

The permeability characteristics of oocyte nuclei are quite different. Figs. 7 and 8 illustrate results of experiments performed on frog oocytes which are the counterparts of those of Figs. 2 and 4 performed on *Drosophila* gland cell nuclei. The oocyte nuclear membrane has so low a resistance that it is indistinguishable from that of the nucleoplasm and cytoplasm; and there is no detectable resting potential across the membrane (Table 1). As far as its electrical characteristics are concerned, the oocyte nucleus behaves merely like a small droplet of nucleoplasm without the additional surface resistance of a membrane. This applies to oocyte nuclei of all phyla heretofore examined, including vertebrates, echinoderms, annelids, molluscs, coelenterates (Table I) (LOEWENSTEIN and KANNO 1963 a; KANNO and LOEWENSTEIN 1963; ASHMAN, KANNO and LOEWENSTEIN, unpublished).

Permeability and Membrane Structure

The elementary membrane structure of the nucleus of *Drosophila* gland cells, as seen in electron micrographs, is a double membrane interrupted at frequent intervals (0.1μ) by gaps of 0.05μ in diameter (GAY 1956; WIENER, SPIRO and LOEWENSTEIN, unpublished data) (Fig. 9). The finding that the membrane offers a high resistance to the passage of ionic currents

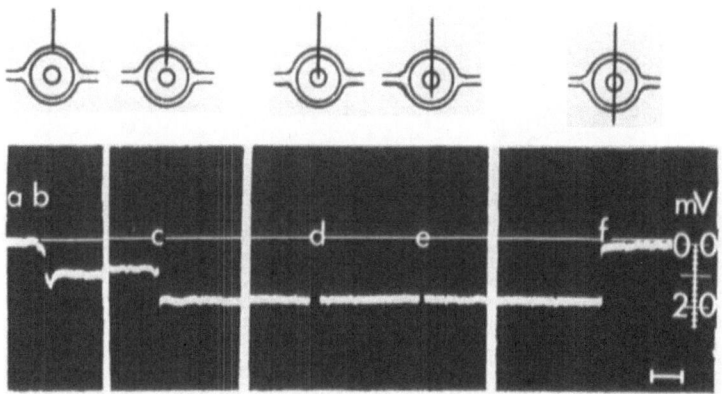

Fig. 7. Potentials across the cell and nuclear membrane of an oocyte (*Xenopus laevis*). Experimental procedure and sequence as in Fig. 2, except for *b* when electrode penetrates through the epithelial cover of the oocyte. Recording microelectrode is *a*, in external fluid; *b*, entering epithelial layer; *c*, entering cytoplasm; *d*, entering nucleoplasm; *e*, leaving nucleus and entering again into cytoplasm; *f*, emerges into the exterior. Note the absence of a nuclear membrane potential. Time calibration: 2 sec. (From KANNO and LOEWENSTEIN, 1963.) Reprinted by permission of Experimental Cell Research.

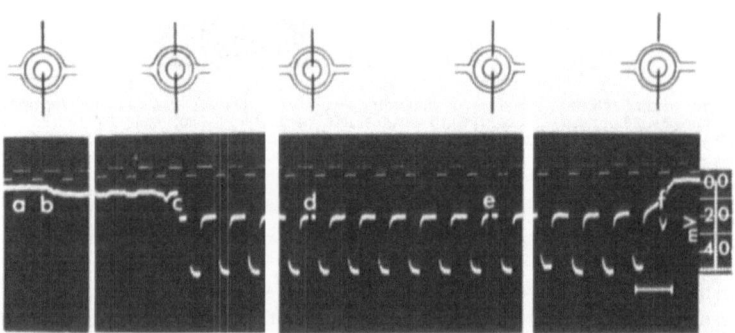

Fig. 8. Resistance of cell and nuclear membranes (oocyte *Xenopus laevis*). Procedure and sequence as in Fig. 4. *a*, recording electrode in cell exterior; *b*, penetrating the epithelial cover; *c*, penetrating the cell membrane; *d*, penetrating the nuclear membrane; *e*, leaving the nuclear membrane (dimming of beam signals entry and exit); *f*, leaving the cell membrane and epithelial cover. Time: 200 msec. (From KANNO and LOEWENSTEIN, 1963.) Reprinted by permission of Experimental Cell Research.

sheds some light on the problem of whether these gaps offer direct connections between nucleoplasm and cytoplasm. (For a discussion of the structural aspects of this problem, see GALL's chapter, pp. 4). The resistance of the nuclear membrane is about one hundredth of that of the surface membrane of the cell. It is also smaller than that of the surface

membrane of a wide variety of other cells. But it is still large enough to
indicate that the nuclear membrane must be a formidable diffusion barrier.
To illustrate this point, consider a layer of nucleoplasm of the thickness

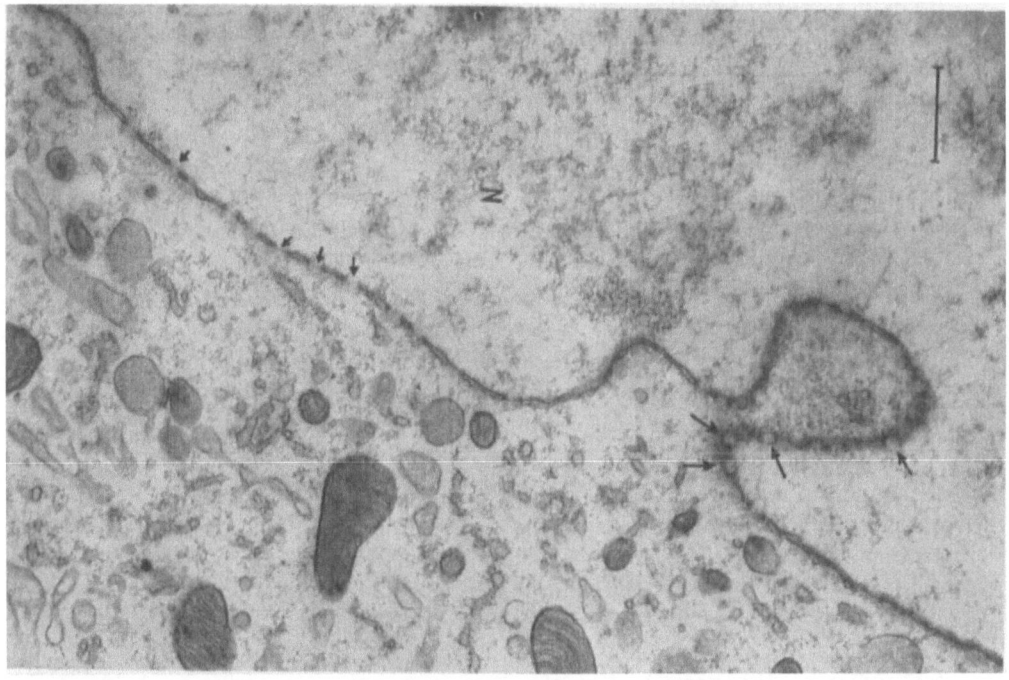

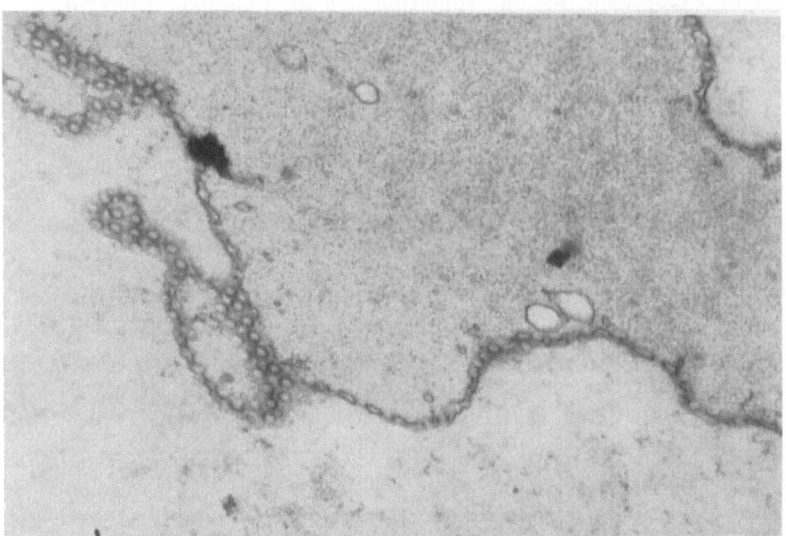

Fig. 9. Electronmicrographs of nuclear membranes of a *Drosophila flavorepleta* gland cell (top) and a *Xenopus laevis* oocyte (bottom). Gaps in membrane structure in transverse (short arrows) and oblique sections (long arrows). Material fixed in OsO_4. Calibration: 0.5 μ. (from J. WIENER. D. SPIRO and W. R. LOEWENSTEIN, unpublished).

of the nuclear membrane (100–200 Å). Such a membrane made of nucleoplasm (specific resistance 100 Ω cm.) would offer a resistance of 10^{-4} Ω cm.^{2}, 4 orders of magnitude lower than the resistance of the actual nuclear membrane. Thus the product of ionic mobility and ionic concentration in the nuclear membrane is one ten-thousandth of that of the nucleoplasm. This obviously does not fit the idea of free membrane pores of the size and frequency mentioned above.

One may now also calculate the resistance of a "porous" membrane and check it against the measured membrane resistance. On the basis of the gap distribution and diameter given by electron micrographs, it is calculated that a "porous" membrane would have a resistance of the order of $10^{-3} \, \Omega \, cm.^2$, three orders of magnitude smaller than the observed membrane resistance in gland cell nuclei (Loewenstein and Kanno 1963 b). Clearly, here, the gaps in membrane structure cannot offer direct contact between nucleoplasm and cytoplasm.

It is interesting in this connection that recent work with labelled elements also reveals the nuclear membrane as a strong diffusion barrier. For example, Allfrey et al. (1961) have shown that the entry of amino acids into the nucleus is not a simple diffusion process, but involves a mechanism of specific transport.

What structural elements of the nuclear surface account for the high resistance? There are no reasons to doubt that the two unit membranes are actually interrupted at the region where they fuse together; transverse sections through the nuclear envelope of a wide variety of cells give evidence for this (Bahr and Beermann 1954; Gall 1954; Afzelius 1955; Palay and Palade 1955; Watson 1955, 1959; Gay 1956; Pappas 1956; Wischnitzer 1958; Robertson 1959; Merriam 1961; for additional references see pp. 5 of this series). But the gap may not necessarily be bridged by nucleoplasm or cytoplasm, as has sometimes been assumed. Electron micrographs of thin sections of the gap space of the nuclei studied with electrical techniques reveal, in fact, the presence of electrondense material, more diffuse than the unit membranes, in most of the gaps (Fig. 9). Formations of this and other kind have been seen also in other nuclei (G. Palade 1963; Afzelius 1955; Watson 1955; Wischnitzer 1956; Merriam 1961; see also pp. 14. It is tempting to speculate that these formations are the additional diffusion barriers which confer upon the nuclear envelope its high electrical resistance.

There is nothing to show as yet at the electron microscope level that would explain the marked disparity in permeability between the membranes of the gland cell and oocyte nuclei. It is disappointing that with all the resolving power attained in the structural and electrophysiological approaches of investigation, there is still such a noticeable lack of correlation between the two. One of the reasons is probably that the features most strikingly revealed by so widely differing techniques do not necessarily relate to the same phenomena. Under the electron microscope the large gap in membrane structure may be the most conspicuous detail; while in terms of permeability, a fine channel in the material bridging the gap, which may easily be overlooked or discarded as an artifact, may be the significant thing.

References

AFZELIUS, A., 1955: The ultrastructure of the nuclear membrane of the sea urchin oocyte as studied with the electron microscope. Exper. Cell Res. **8**, 147.

ALLFREY, V. G., R. MEUDT, J. W. HOPKINS, and A. E. MIRSKY, 1961: Sodium-dependent "transport" reactions in the cell nucleus and their role in protein and nucleic acid synthesis. Proc. Nat. Acad. Sci., Washington, **47**, 907.

BAHR, G. F., and W. BEERMANN, 1954: The fine structure of the nuclear membrane in the larval salivary gland and midgut of *Chironomus*. Exper. Cell Res. **6**, 519.

CALLAN, H. G., and S. G. TOMLIN, 1950: Experimental studies on amphibian oocyte nuclei. I. Investigation of the structure of the nucleus membrane by means of the electron microscope. Proc. Roy. Soc. Lond. B. **137**, 367.

CHAMBERS, R., and H. B. FELL, 1931: Micro-operation on cells in tissue culture. Proc. Roy. Soc. Lond. B. **109**, 380.

COLE, K. S., 1940: Permeability and impermeability of cell membranes for ion. Cold Spring Harb. Symp. Quant. Biol. **8**, 110.

ECCLES. J. G., 1954: Neurophysiological Basis of Mind. Oxford Univ. Press, Oxford 1953.

GALL, J. G., 1954: Observations on the nuclear membrane with the electron microscope. Exper. Cell Res. **7**, 197.

GAY, H., 1956: Chromosome-nuclear membrane-cytoplasmic interrelations in *Drosophila*. J. Biophys. Biochem. Cytol. **2** (suppl.), 407.

HARTMANN, J. F., 1952: Electron microscopy of nuclei in nerve cells. Anat. Rec. **112**, 340.

KANNO, Y., and W. R. LOEWENSTEIN, 1963: A study of the nucleus and cell membranes of oocytes with an intra-cellular electrode. Exper. Cell Res. **31**, 149.

KOPAC, M. J., and G. M. MATEYKO, 1958: Malignant nucleoli in cytological studies and perspectives. Ann. N. Y. Acad. Sci. **73**, 237.

LOEWENSTEIN, W. R., and Y. KANNO, 1962: Some electrical properties of the membrane of a cell nucleus. Nature **195**, 462.

— — 1963 a: The electrical conductance and potential across the membrane of some cell nuclei. J. Cell Biol. **16**. 421.

— — 1963 b: Some electrical properties of a nuclear membrane examined with a microelectrode. J. Gen. Physiol. **46**. 1123.

MERRIAM, R. W., 1961: On the fine structure and composition of the nuclear envelope. J. Biophys. Biochem. Cytol. **11**. 559.

MIRSKY, A. E., and S. OSAWA, 1961: The interphase nucleus. In: The Cell II (Brachet and Mirsky, editors). New York, Academic Press.

PALADE, G., Personal communication.

PALAY, S. L., and G. E. PALADE, 1955: The fine structure of neurons. J. Biophys. Biochem. Cytol. **1**, 69.

PAPPAS. G. D., 1956: Fine structure of the nuclear envelope of *Amoeba proteus*. J. Biophys. Biochem. Cytol. **2** (suppl.), 431.

ROBERTSON, J. D., 1959: Ultrastructure of cell membranes and their derivatives. Biochemical Soc. Symposium **16**, 3.

WATSON, M. L., 1955: The nuclear envelope. its structure and relation to cytoplasmic membranes. J. Biophys. Biochem. Cytol. **1**, 257.

— 1959: Further observations on the nuclear envelope of the animal cell. J. Biophys. Biochem. Cytol. **6**, 147.

WIENER. J. D. SPIRO. and W. R. LOEWENSTEIN. 1964: J. Cell Biol. (in press).

WISCHNITZER, J.. 1958: Electron microscope study of the nucleus envelope of amphibian oocytes. J. Ultrastr. Res. **1** (3). 251.

The Permeability Characteristics of the Nuclear Envelope at Interphase

By

C. M. FELDHERR and C. V. HARDING

Department of Physiology, University of Alberta, Edmonton; and the Departments of Ophthalmology and Physiology, College of Physicians and Surgeons, Columbia University, New York

With 4 Figures

Contents

I. Introduction

It is well known that a variety of essential cellular processes are dependent either wholly, or in part, on the cell nucleus. Similarly, the nucleus, in order to carry out its functions, is dependent on the cytoplasm (BRACHET 1957; MIRSKY and OSAWA 1961). The materials which are responsible for these interactions must, in some way, cross the nuclear envelope. For this reason, knowledge of the permeability characteristics of the nuclear envelope can be of considerable value in understanding the interactions and exchanges that take place between the nucleus and the cytoplasm. For example, if the permeability characteristics of the nuclear envelope were fully understood, one should know in what specific form and by what mechanism RNA can pass from the nucleus into the cytoplasm, what substances, due to their inability to cross the nuclear envelope, would have to be synthesized in the region of the cell where they are located. Furthermore, as the mechanisms of such exchanges become known, it may be

3*

possible to alter the interactions between the nucleus and the cytoplasm and thus control various physiological processes.

At present, little is known regarding the degree to which the nuclear envelope regulates nucleocytoplasmic exchanges, however, an attempt has been made to collect and evaluate some of the available information concerning the types of material that can cross the nuclear envelope, and the various pathways and possible mechanisms involved in such exchanges.

II. Exchanges Across the Nuclear Envelope

A. General Considerations

There are basically three approaches which have been used to study transfer across the nuclear envelope. These include morphological studies, performed with the electron microscope, permeability studies on isolated nuclei, and permeability studies on nuclei within intact cells.

Morphological studies have been of considerable value for two reasons; first, it has been possible by means of electron microscopy to describe the structure of the nuclear envelope; second, from its structure and relationships with other cell components, various pathways for nucleocytoplasmic exchanges can be deduced.

To determine whether a specific substance or group of substances can penetrate the nuclear envelope, one cannot depend on morphological studies alone; experiments to determine the permeability characteristics of the nuclear envelope must be conducted. Such experiments have been done with isolated nuclei and nuclei within intact cells. Experimentation with isolated nuclei is a much simpler approach. Unfortunately, when using such material, one is faced with the possibility that the permeability characteristics are altered as a result of the isolation procedures. For this reason, experiments on nuclei within the intact cell are desirable and will be given more emphasis in the following discussion.

A large majority of the permeability studies on nuclei have been concerned with the passage of material into the nucleus, rather than out of the nucleus. The mechanism by which a given substance leaves the nucleus may possibly differ from the mechanism by which it enters. Further experimentations with the techniques now available should resolve this point.

B. Inorganic Salts

When isolated nuclei from frog oocytes are exposed to various ions, structural or colloidal alterations, which involve opacity changes, often occur within the nucleus. DURYEE (1940) interpreted such changes to mean that the ions producing alterations within the nucleus have penetrated the nuclear envelope. From his studies on anions, he concluded that hydroxide, sulphide, iodide, thiocyanate, citrate, acetate, sulfate and chloride can penetrate the nuclear envelope. The cations which entered the nucleus included potassium, sodium, magnesium and calcium. From the rate at which changes occurred within the nuclei, DURYEE further concluded that anions penetrate more rapidly than cations. However, as has been pointed out by CALLAN (1952), the evidence for the latter conclusion seems inadequate. ANDERSON and WILBUR (1952) studied the effects of inorganic salts on isolated liver nuclei. The structural changes observed indicated that the nuclear envelope is permeable to NaCl, KCl, $MgCl_2$ and $CaCl_2$.

Information concerning the passage of salts (as well as other substances to be discussed later) across the nuclear membrane has been obtained by using osmotic methods. It has been found by various workers that isolated oocyte nuclei will swell when placed in a variety of inorganic salt solutions (e. g. GOLDSTEIN and HARDING 1950; CALLAN 1949; BATTIN 1959). It has generally been concluded, from these experiments, that the salts readily enter the nucleus and that swelling is caused by the presence within the nuclear sap of osmotically active substances that cannot leave the nucleus. This explanation has been questioned by HUNTER and HUNTER (1961). They observed that the rate of swelling of isolated frog oocyte nuclei decreased as the concentration of NaCl in the medium was increased. Similar results were obtained with glycerol. These findings, according to HUNTER and HUNTER, are not consistent with osmotic principles. In interpreting such results, one must consider the possibility that the substances tested may have affected the permeability characteristics of the nuclear envelope (CALLAN 1952) and, furthermore, that the extent of the effects may have depended on the concentration of the test material. If such were the case, osmotically active substances might, for example, leave the nucleus more rapidly in high concentrations of NaCl than at lower concentrations of the same salt. There is, in fact, experimental evidence that sodium does influence the movement of materials across the nuclear envelope (see later section). Under these conditions, the interpretation, based on osmotic principles, would not be inconsistent with the results of HUNTER and HUNTER. Until this point is tested further it would seem unjustified to disregard all the osmotic studies that have so far been performed.

Although, at present, the osmotic approach seems to be of value in studying oocyte nuclei, it is likely that certain isolated somatic cell nuclei either do not react osmotically, or that such changes are obscured by changes in hydration of the chromosomes (ANDERSON and WILBUR 1952).

In addition to the studies on isolated nuclei, investigations of the permeability of the nuclear envelope to ions have been carried out using intact cells. From the changes in nuclear volume which were observed after micro-injecting frog oocytes with a solution containing KCl and NaCl, it was concluded that these salts readily penetrate the nuclear envelope (HARDING and FELDHERR 1959). ABELSON and DURYEE (1949) treated intact frog oocytes with radioactive sodium. Autoradiograms showed that the sodium had not only entered the nuclei, but had concentrated there. NAORA et al. (1962) confirmed the above results and, using radioactive tracers, also demonstrated that phosphate, sulfate and potassium concentrate in the nuclei. One explanation put forth for the greater content of certain ions in the nucleus is that the water content in oocyte nuclei is higher than that of the cytoplasm (ABELSON and DURYEE 1959). However, the results of NAORA et al. (1962), which were corrected for the water contents of nucleus and cytoplasm, still show a greater concentration in the nucleus than the cytoplasm. It is possible that, after entering the nucleus, the ions are adsorbed to substances within the nucleoplasm and thus removed from solution.

Recently LOEWENSTEIN and KANNO (1963), using microelectrodes, measured the resistance across the nuclear envelope in intact oocytes and gland cells. The results obtained for oocytes are consistent with previous findings, since they show that ions can readily penetrate the nuclear envelope. The resistance (1.5 $\Omega \, cm^2$) across the nuclear envelope of *Drosophila* salivary gland cells, however, shows that the nuclear envelope, in this instance, acts as a barrier to the diffusion of ions. This finding is highly significant with regard to the structure of the nuclear envelope and will be considered in a later section. The results obtained with *Drosophila* salivary gland cells also point out the possibility of important differences in permeability between somatic cell nuclei and germinal vesicles. Differences in permeability among the nuclei from different cell types or in the same cell under different physiological conditions might also be expected. This perhaps should be kept in mind in considering the varibilty of the results obtained.

Generally, the results obtained to date indicate that ions can enter both isolated nuclei and nuclei within intact cells. Further experiments for the purpose of determining the rate of penetration of various anions and cations, as well as the mechanisms involved, would be of interest.

C. Low Molecular Weight Molecules

Using isolated amphibian oocytes, it has been found by osmotic methods that sugars such as xylose, glucose, sucrose and raffinose can rapidly penetrate the nuclear envelope (GOLDSTEIN and HARDING 1950; CALLAN 1952).

As has been pointed out by MIRSKY and OSAWA (1961), numerous experiments involving the use of isotopically labelled amino acids have shown that such substances can penetrate the nuclear envelope. This would also hold for various nitrogenous bases, including labelled thymidine and cytosine, used to study DNA and RNA synthesis. Experiments with intact amphibian oocytes have indicated that C^{14}-leucine and C^{14}-alanine not only enter but concentrate in the nuclei (NAORA et al. 1962).

By micro-injecting dyes near the nucleus of intact amoebae, it is possible to study the passage of these substances across the nuclear envelope. The presence of color within the nucleus indicates that penetration of the dye has occurred.

Such experiments were carried out by MONNÉ (1935), who found that acid or basic dyes, lipoid soluble or lipoid insoluble dyes, crystalloidal or colloidal dyes and organic or inorganic dyes all penetrated the nuclear envelope. CLARK (1943), using the same techniques was able to verify these results.

As in the case of inorganic salts, there is considerable evidence, from both isolated nuclei and nuclei within intact cells, that other substances having molecular weights of approximately 500 or less can cross the nuclear envelope.

D. Macromolecules

A number of investigations have been performed to study the osmotic properties of nuclei within intact cells (SHINKE 1937; BUCK and BOCHE 1938; HAMBURGER 1904; CHURNEY 1942; BECK and SHAPIRO 1936). These studies involved placing whole cells in solutions of different salt concentrations. It was found, for a variety of cell types, that under such conditions the nucleus undergoes persisting changes in volume which in many instances suggest that it behaves as an osmometer. It can be concluded from these experiments that the nuclear envelope is semipermeable and that there are some soluble materials present within the nucleus and cytoplasm which do not freely cross this structure. Since it is doubtful that such materials are either ions or low molecular weight molecules (see previous sections), one must consider whether or not macromolecules can penetrate the nuclear envelope. Such knowledge is of particular significance in view of the possibility that information from the nucleus to the cytoplasm may be carried in the form of high molecular weight RNA molecules.

Much of the work on the passage of macromolecules across the nuclear envelope has been done on isolated nuclei. Using osmotic methods, it has been concluded that the nuclear envelope of amphibian oocytes is relatively impermeable to egg albumen (GOLDSTEIN and HARDING 1950; CALLAN 1949), glycogen (CALLAN 1949), acacia (CALLAN 1949), and bovine serum albumin (BATTIN 1959). HOLTFRETER's experiments (1954), on the other hand, indicate that the nuclei of amphibian oocytes are permeable to hemoglobin. MERRIAM (1959 b), using interference microscopy, provided evidence that bovine serum albumin can enter nuclei isolated from *Chaetopterus* eggs; further-more, he found that soluble elements can readily leave the nucleus. STERN and MIRSKY (1953) found that proteins are lost from thymus and liver nuclei when they are isolated in aqueous media. ANDERSON and WILBUR (1951), using isolated liver nuclei, concluded that heparin can penetrate the nuclear envelope. In a later investigation, ANDERSON (1953) studied the effects of DNAase, RNAase, trypsin and chymotrypsin on isolated liver nuclei. His results suggested that these substances can penetrate the nuclei. In ex-periments such as these, however, the possible effects of the enzymes on the structure of the nuclear envelope could cause complications.

There are numerous other studies that have been carried out on isolated nuclei, but the experiments just referred to indicate the variability of the results. Because of this, it is difficult to draw any definite conclusions regarding the permeability of the nuclear envelope in the intact cell from such studies on isolated nuclei. The reason for the variability of the results is not clear, but two likely explanations are:

1. that, during isolation, the permeability characteristics of the nuclear envelope may be altered due to irreversible chemical changes;

2. that the permeability characteristics, although not basically altered, may vary depending on the nature of the isolation medium. Whatever the actual reason, it would be preferable to study nuclei within intact cells. Such experiments have been carried out and will now be considered.

Many cells can take up protein that has been introduced into the extracellular medium. The later distribution of the protein within the cells can be determined by using fluorescein-labelled antibodies (to stain unlabelled antigen) or protein labelled directly with fluorescein or other markers. The results of such experiments, which have been carried out mostly on somatic cells, have been considered in some detail by HOLTZER and HOLTZER (1960) and will not be discussed individually here. It should be pointed out, however, that the findings, in regard to the ability of protein to penetrate the nuclear envelope, are quite contradictory. The reason for this could be, according to HOLTZER and HOLTZER, a technical one. They believe that cells in which protein has been found to enter the nucleus were injured. Their own experiments are in agreement with this view. It is also conceivable that proteins which have been introduced by way of the extracellular medium, and therefore, most likely incorporated in pinocytosis droplets (NACHMIUS and MARSHALL 1961), are, under normal conditions, not available to the nucleus.

The proteins used in the labelling experiments described above were primarily acidic. Basic proteins, when added to the extracellular medium appear to have less difficulty entering the cells, and in many instances seem to alter, structurally and chemically, both the cytoplasm and the nucleoplasm. On the basis of such changes, BECKER and GREEN (1960), concluded that histones and protamines can penetrate the cell membrane and the nuclear envelope of ascites cells. FISCHER and WAGNER (1954) also found that these substances can penetrate the nuclei of intact cells. Since the cells are seriously injured following treatment with histones and protamines (BECKER and GREEN 1960), it has been suggested by MIRSKY and OSAWA (1961) "... that penetration may actually have been made possible by the damage caused." BRACHET (1955; 1956) observed that the nuclei of both amoebae and onion root cells are modified following treatment of whole cells with ribonuclease, and concluded that this protein can penetrate the nuclear envelope. A similar conclusion could be drawn from the work of FICA and ERRERA (1955) who found that ribonuclease affected the uptake of labelled amino acids in the cytoplasm, nucleus and nucleoli of oocytes. The exact meaning of the results obtained with ribonuclease in regard to the permeability of the nuclear envelope might be questioned on the basis that some RNA-containing component of the nuclear envelope may be digested by the enzyme, thus altering its characteristics. Since the effects of ribonuclease are reversible (FICA and ERRERA 1955), this criticism is probably not a very serious one. It is perhaps also conceivable that the normal metabolism of the cytoplasm is so altered by exposure to ribonuclease that secondary effects on the synthetic activities of the nucleus result.

By combining the techniques of autoradiography and nuclear transplantation, it has been demonstrated by GOLDSTEIN and PLAUT (1955) that

RNA or a precursor of RNA passes from the nucleus to the cytoplasm in amoebae. In later experiments, GOLDSTEIN (1958) found that a substance, considered to be a protein, normally passes both in and out of the nuclei, and even concentrates in the nuclei. More recently, BYERS, PLATT and GOLDSTEIN (1963) have provided convincing evidence that this substance is indeed protein in nature. The value of such experiments, in determining the types of the substance actually involved in nucleocytoplasmic exchanges is obvious.

In experiments designed specifically to study the permeability characteristics of the nuclear envelope to macromolecules, HARDING and FELDHERR (1959) microinjected various concentrations of non-toxis polyvinylpyrrolidone (PVP) (M. W. approx. 40,000) and bovin serum albium (M. W. approx. 65,000) directly into the ground cytoplasm of intact frog oocytes. The changes in nuclear volume which were observed three minutes after injection were related to the colloid osmotic pressure of the injected solutions (Fig. 1). It was concluded, therefore, that molecules of 40,000 molecular weight or larger, do not freely penetrate the nuclear envelope in oocytes. Similar results were obtained by FELDHERR and FELDHERR (1960), who demonstrated that fluorescein-labelled gamma globulin did not

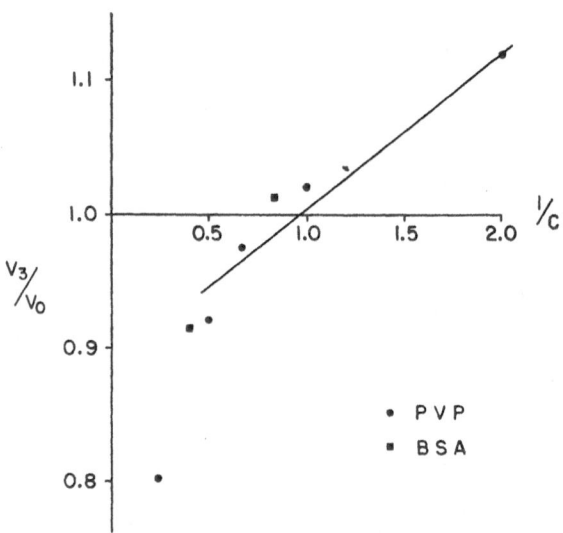

Fig. 1. The ratio of nuclear volume at 3 minutes after injection, over the volume before injection, is plotted as a function of the inverse of the per cent concentration of PVP injected. The circles represent PVP and the squares, BSA. The per cent concentrations of BSA are expressed in terms of the per cent concentrations of PVP which have equivalent molarities. The straight line is a regression calculated on the basis of all PVP experiments except those in which 4 per cent PVP was used. Reproduced with permission of the journal of general physiology.

enter the nuclei of *Cecropia* oocytes within ten minutes after being injected into the ground cytoplasm of these cells. It should be emphasized, that inability of macromolecules to penetrate the nuclear envelope in short-term experiments does not rule out the possibility that over longer periods, macromolecules can enter the nucleus. Furthermore, the methods employed were not sensitive enough to detect individual molecules, and thus it is not unlikely that some molecules had penetrated the nucleus, but were not observed.

To determine whether macromolecules can penetrate the nuclear envelope over long time intervals, FELDHERR (1962 b) microinjected ferritin into the ground cytoplasm of multinucleated amoebae and determined its later distribution by means of electron microscopy. Since individual ferritin molecules can be detected with the electron microscope, the sensitivity of this method, in comparison with the previous studies is considerably greater. The results of these experiments indicated that ferritin (M. W. approx. 550,000; diameter approx. 95 Å; isoelectric point 4.4) can penetrate the nuclear envelope (Fig. 2). The protein was consistently found in the nuclei one hour and 24 hours after injection. Some ferritin was also present in the nuclei after seven minutes, but not consistently. In similar experiments it was shown that colloidal gold particles, coated with PVP (estimated overall size 100–200 Å), can also enter the nuclei, and in fact, are concentrated in these organelles after 24 hours (FELDHERR and MARSHALL 1962). If amoebae are fixed two minutes after being injected with colloidal gold, however, practically no gold can be found in the nuclei (FELDHERR 1962 a). This suggests, as do the studies on oocytes, that particles of macromolecular dimensions do not *freely* penetrate the nuclear envelope.

The ability of colloidal particles to enter the nucleus has also been demonstrated by MOORE et al. (1961), who found colloidal iron particles in tissue cell nuclei after injections into whole animals.

From work on intact cells one could conclude, at present, that although macromolecules cannot freely penetrate the nuclear envelope, they can in some way enter the nucleus. The ability to enter the nucleus does not seem to be strictly a property of charge, since both basic and acidic substances seem able to cross the nuclear envelope. One possible characteristic which could limit penetration is molecular size. Preliminary experiments with colloidal gold (FELDHERR unpubl.) have indicated that there is a size above which the colloidal particles cannot enter the nucleus. This size appears to be about 125–145 Å in diameter, which is smaller than the annuli of the nuclear envelope (see next section).

The evidence that the nuclear envelope has the properties of a semipermeable membrane is not necessarily inconsistent with the evidence that macromolecules can pass between the cytoplasm and the nucleus. In the case of the amphibian oocyte (HARDING and FELDHERR 1959) it has been pointed out that—"the osmotic experiments with germinal vesicles indicate a physical or physiological impermeability to large molecules." It was considered conceivable that—"the penetration of large molecules does indeed occur under conditions in which the nucleus would still maintain itself as an osmometer. Just for example, if a process comparable to pinocytosis should occur at the nuclear surface, cytoplasmic proteins could be taken into the nucleus, while the nuclear membrane itself retained its low permeability to these molecules at all times."

STERN and MIRSKY (1953) were the first to emphasize this point—"it is questionable, however, whether 'semipermeability' of a membrane and the passage of macromolecules across it are incompatible. If a parallel is drawn from the many studies of 'active transport' across cell membranes, it would appear quite probable for the nuclear membrane to act structurally as a barrier to free diffusion between nucleus and cytoplasm while functioning by energy-requiring mechanisms in the transportation of substances to and from the nucleus."

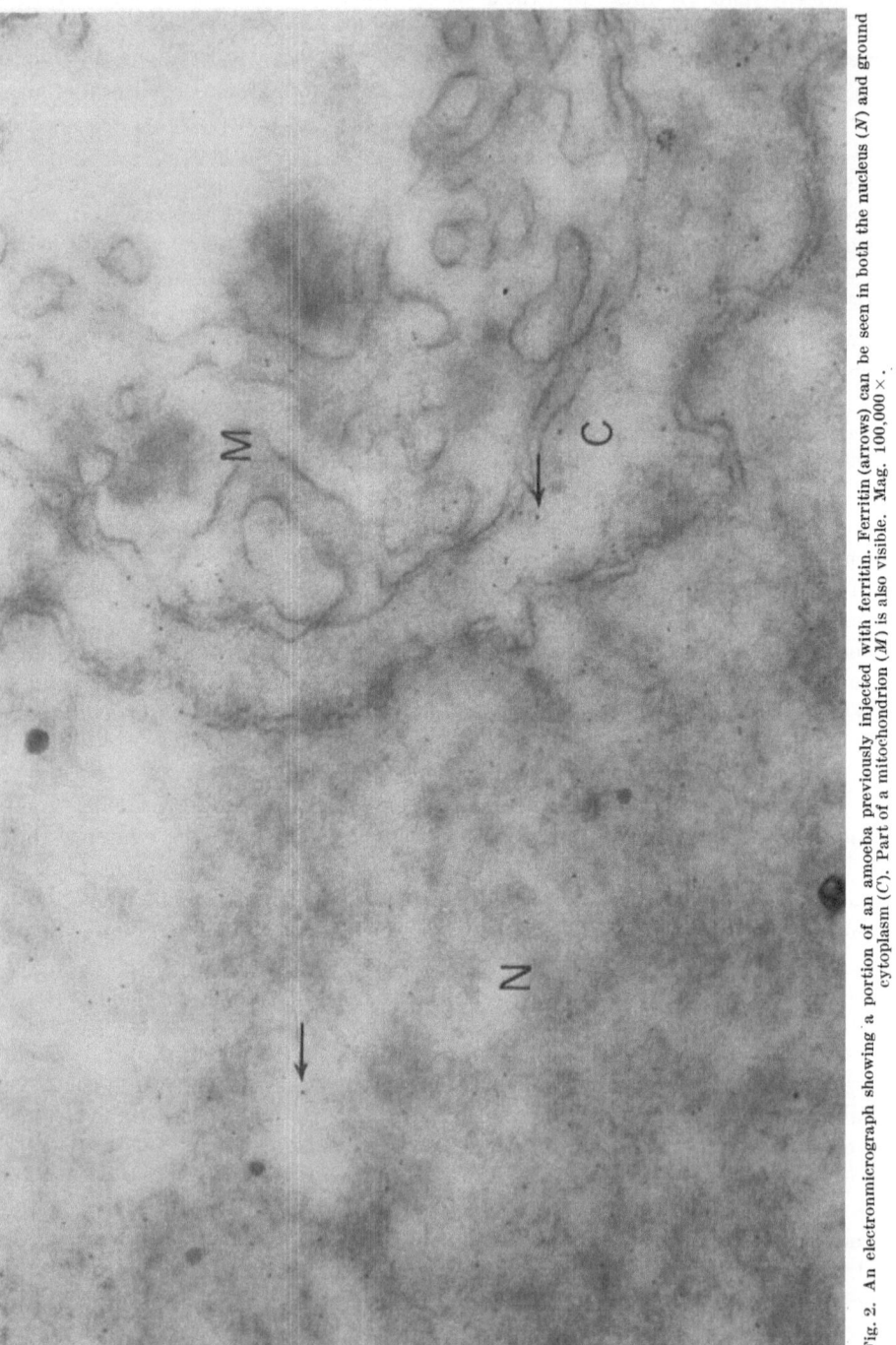

Fig. 2. An electronmicrograph showing a portion of an amoeba previously injected with ferritin. Ferritin (arrows) can be seen in both the nucleus (*N*) and ground cytoplasm (*C*). Part of a mitochondrion (*M*) is also visible. Mag. 100,000 ×.

III. Possible Pathways and Mechanisms
of Nucleocytoplasmic Exchanges

Having established that various substances, including at least some macromolecules, are capable of penetrating the nuclear envelope, one must inquire as to the pathways and mechanisms of such exchanges. Before discussing what information is available, it would be advantageous to examine first the structure of the nuclear envelope.

In the light microscope, the nuclear membrane appears as a barrier between the nucleoplasm and cytoplasm, but no detailed structures can be detected. Studies of the ultrastructure of the nuclear envelope with the electron microscope, however, have been extremely profitable. There is a characteristic structure that has been observed in all cell types studied. The observation of this structure with a variety of fixatives and embedding methods makes it unlikely that the findings can be attributed to artifacts resulting from the preparation of the material.

Typically the nuclear envelope, as seen with the electron microscope, consists of two membranes, separated by a space (the perinuclear space) of about 150 Å. In perpendicular sections the membranes are, in certain areas, discontinuous. These discontinuous regions are often referred to as pores. In tangential sections the discontinuities appear as annuli which can be as large as 1000 Å in diameter. The structure of the nuclear envelope as described thus far, is well established (although variations have been reported, e. g. Pappas 1956); however, there is some controversy as to the specific nature of the pores. In some instances the pores seem to represent areas where the cytoplasm is in direct contact with the nucleoplasm, in other instances there appears to be a thin membrane across the pores. Centrally located granules, as well as tube-like elements have also been found in association with these areas. Further details and references regarding the ultrastructure of the nuclear envelope can be obtained from recent reviews by Wischnitzer (1960), Baud (1959) and Gall (1963).

From a structural standpoint, the pores of the nuclear envelope would appear to be likely pathways for nucleocytoplasmic exchanges. This possibility has, in fact, been suggested by numerous investigators (e. g. Watson 1955; Whaley et al. 1960). Evidence in support of this view has been provided by Pollister et al. (1954), who described filamentous material extending through the pores in oocytes. Anderson and Beams (1956) studying nurse cell nuclei also observed electron-opaque material extending through the pores. Feldherr (1962 a) found that colloidal gold particles coated with PVP, were frequently located within the pores one and two minutes after being injected into the cytoplasm of amoebae (Fig. 3). Furthermore, the gold particles were located specifically in the centers of the pores. Since at later time intervals the colloidal particles were found in the nuclei, it was concluded that passage across the nuclear envelope occurs, at least in part, through the pores. The simplest explanation for the fact that the particles were located specifically in the centers of the pores is that within each pore, a thick-walled tube exists, which restricts passage to a central channel. As has been pointed out, there is also morphological evidence which supports this view (WISCHNITZER 1958; AFZELIUS 1955; WATSON 1959). FELDHERR's work also provides a clue as to the nature of the granules which are normally located in the centers of the annuli in a variety of cells (e. g. WISCHNITZER 1958; WATSON 1959; GAY 1956). One would suspect that such granules, rather than being part of the structure of the pores, are in the process of penetration. It was also suggested on the basis

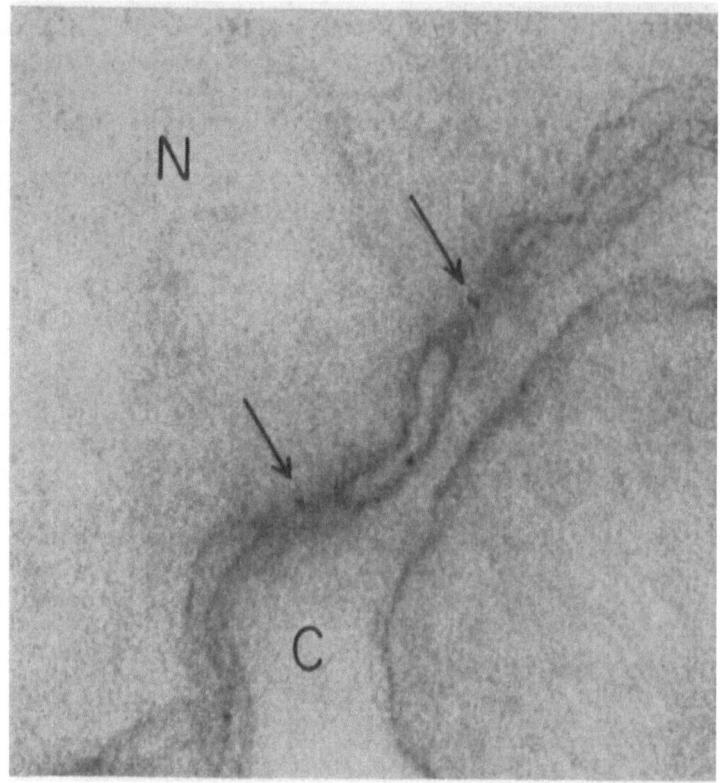

Fig. 3. A section through an amoeba, fixed one minute after being injected with PVP-coated colloidal gold. Two pores can be seen, both containing gold particles (arrows). *N*, nucleus; *C*, cytoplasm. Mag. 216,000 ×

that gold particles were so frequently found in these areas, that binding of material to some component of the pores may occur. To investigate further the possibility that substances bind to the pore material (that electron-dense material frequently seen within and associated with the pores), experiments were performed on nuclei isolated from frog oocytes (FELDHERR 1963). The nuclei were treated for five minutes with colloidal gold particles coated with either PVP or poly-L-proline. Subsequent examination of the nuclei with the electron microscope showed that gold had accumulated within and adjacent to the pores, but few particles had entered the nuclei (Fig. 4). The accumulation of the gold in these areas and particularly the association of the gold with the electron-dense material present in the pores, provides direct evidence that the pore material can act as a barrier

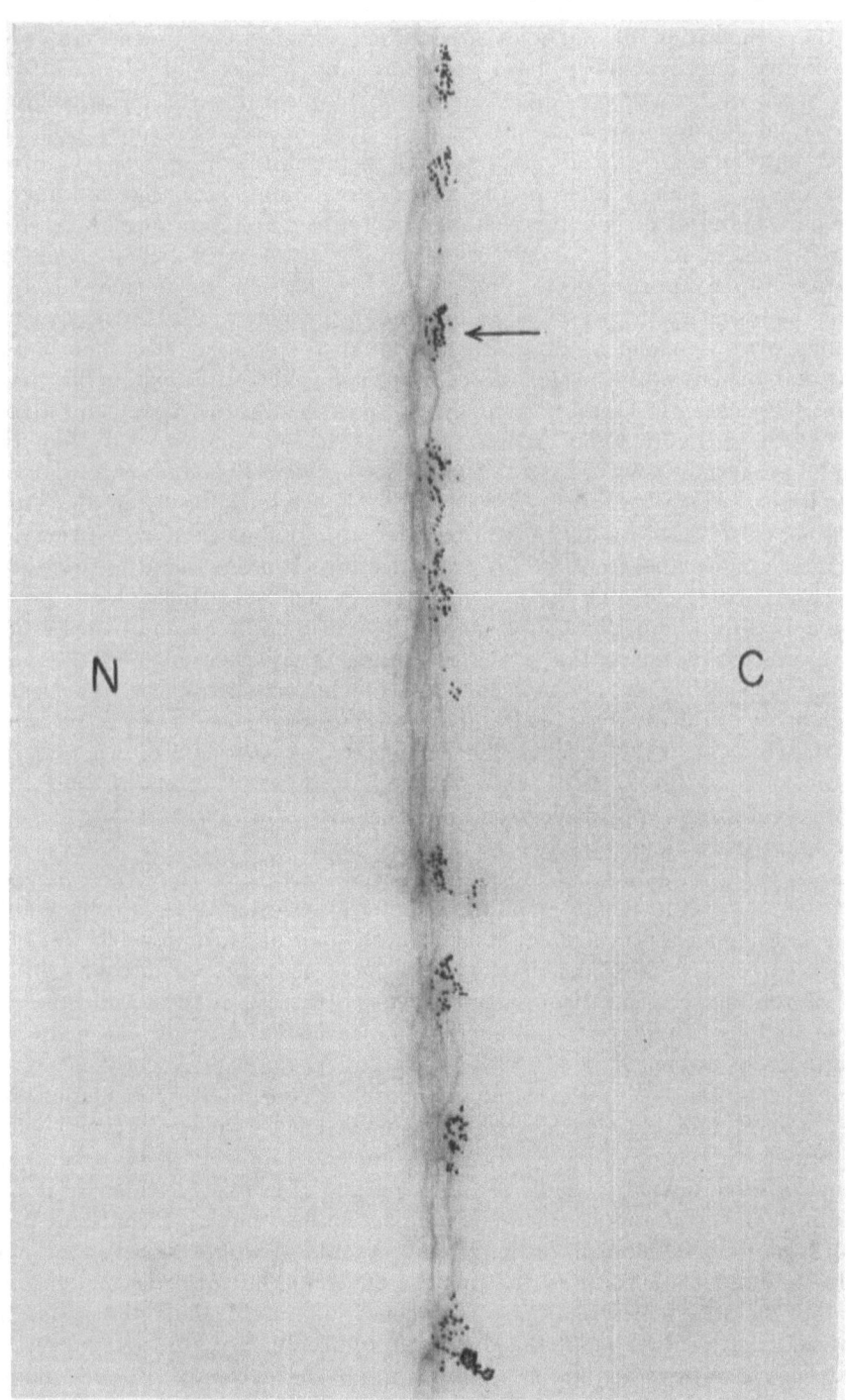

Fig. 4. A section through an isolated nucleus treated with poly-l-proline coated gold particles. The gold (arrow) has accumulated within and adjacent to the pores. *C*, cytoplasmic side of the nuclear envelope; *N*, nuclear side of the envelope. Note—gold extends toward cytoplasmic side of membrane. Mag. 96,000×. Reproduced with permission of the journal of cell biology.

to the penetration of particles having macromolecular dimensions. This possibility has previously been considered by MIRSKY and OSAWA (1961). The work of LOEWENSTEIN and KANNO (1963) concerning the electrical properties of the nuclear envelope suggests that in salivary gland cells the pore material may also act as a barrier to the diffusion of ions. The fact that the gold accumulates in the pores further indicates that binding to the pore material does occur. Although the exact function of such binding is not known, it is of interest to speculate as to its possible effects on nucleocytoplasmic exchanges. One of the first phenomena to come to mind when considering the significance of binding is pinocytosis. An early step in this process is the binding of material to the cell surface. The bound material then stimulates membrane flow and is incorporated in the pinocytosis vesicles which result from the induced membrane movements (HOLTER 1959), NACHMIUS and MARSHALL (1961) have demonstrated that the binding characteristics of the coat substance on the cell surface can function, to a limited extent in selecting the type of molecules to be taken up. Thus, in pinocytosis, binding not only stimulates uptake, but can also concentrate and select molecules from the external medium. Binding could have similar functions with regard to the passage of macromolecules through the pores. The actual uptake of the bound material could then be accomplished either by a process of adsorption and desorption, or by movement of the pore material, resulting in the movement of the bound substances. The latter mechanism could involve membrane flow, as proposed by BENNETT (1956).

MOORE et al. (1961) believe that material may pass directly across the membranes of the nuclear envelope. They injected colloidal iron into rabbits and, by means of electron microscopy, studied its later distribution. Masses of the colloidal particles were occasionally found next to the nuclei of somatic cells, and particles could be seen extending from these masses directly across the membranes of the nuclear envelope. It is conceivable that results of this nature could have been due, at least in part, to knife artifact, i. e. the actual distribution may have been altered during sectioning of the material for electron microscopy. In view of the significance of these findings, further studies to determine the validity of this criticism would be of value.

The continuation of the outer membrane of the nuclear envelope with the endoplasmic reticulum was first observed by WATSON (1955) and represents another possible pathway for nucleocytoplasmic exchanges. This structural relationship would seem to permit free exchange between the perinuclear space and the endoplasmic reticulum; however, material passing from the nucleoplasm to the ground cytoplasm would have to cross not only the inner membrane of the nuclear envelope, but also the membranes of the endoplasmic reticulum. This condition might limit the types of material that could be exchanged by this route. WATSON, in fact, suggested that such a pathway might be important for the exchange of small molecules, whereas large molecules might pass through the annuli.

The investigations of GAY (1956) on salivary gland cells of *Drosophila* and MOSES (1956) on crayfish spermatids have indicated that submicroscopic blebs forming from the nuclear envelope can serve, to some degree, in nucleocytoplasmic exchanges. These blebs are closely associated with chromosomal material, and it is proposed that following their formation they are detached from the nuclear envelope and pass into the cytoplasm. CLARK (1960) has also found nuclear blebs in pancreatic acinar cells. In these cells, the blebs, due to their rather large size, are throught to correspond to the nuclear extrusions described with the light microsope. The extent to which nuclear blebs are involved in exchanges between the nucleus and cytoplasm is not known. From the investigations conducted so far, nuclear blebs seem to be limited to only a few cell types, and even in such cases, they appear only at certain stages of activity.

Another means of exchange, similar to that just discussed, has been proposed by SWIFT (1956). He, as well as other workers (e. g. AFZELIUS 1955; MERRIAM 1959 a), have described, within the cytoplasm of various cell types, annulate lamellae, which closely resemble the nuclear envelope. Due to this similarity in structure, it has been suggested that the annulate lamelae arise from the nuclear envelope, and may serve in the transfer of material from the nucleus to the cytoplasm. As SWIFT points out, however, this hypothesis is, at present, highly speculative, and requires further investigation.

It has already been mentioned that sodium seems to be involved in the penetration of certain substances across the nuclear envelope. In this regard, it was demonstrated by ALLFREY et al. (1961) using isolated thymus nuclei, and by NAORA et al. (1962) using oocyte nuclei that the entrance of amino acids into the nucleus is affected by the presence of sodium. Furthermore, it was demonstrated (ALLFREY et al. 1961) that the mechanism involved is probably enzymatic. Whether all exchanges across the nuclear envelope are similarly controlled will no doubt be a subject of further investigation. It is conceivable, however, that only certain of the pathways for exchange are enzymatically regulated. For example, exchanges directly across the membranes of the nuclear envelope, or by way of the perinuclear space and endoplasmic reticulum (if these are actually pathways) may depend on an enzymatic mechanism, whereas a different mechanism may be involved in passage across the annuli.

One could conclude that the annuli of the nuclear envelope are likely pathways for nucleocytoplasmic transfers, which are available to all cells. The formation of nuclear blebs, and their later detachment into the cytoplasm also seems to be involved in such exchanges. This pathway, however, is probably restricted to only certain cell types. Other means of transfer that have been suggested include; direct passage through the membranes of the nuclear envelope; exchanges by way of the annulate lamellae, and transfer through the endoplasmic reticulum and perinuclear space. Less is known of the specific mechanisms involved in nucleocytoplasmic exchanges. There is evidence, however, that enzymatic processes may be involved at least in the case of amino acids. Adsorption has been suggested as an early step in exchanges taking place through the annuli, but this point requires further study.

IV. Conclusions

From permeability studies of the nuclear envelope one can, at present, draw the following conclusions concerning the types of nucleocytoplasmic exchanges that can be expected to occur:

1. Ions and small molecules (M. W. approx. 500 or less) can readily enter the amphibian oocyte nucleus. It is quite possible that such materials can likewise leave the nucleus.

2. The evidence at present indicates that at least under certain conditions macromolecules are able to pass from the cytoplasm into the nucleus. This holds for both positively and negatively charged materials. Thus, the idea that some of the proteins present in the nucleus could have been synthesized in the cytoplasm, is consistent with the permeability characteristics of the nuclear envelope. The degree to which macromolecules can leave the nucleus remains to be established.

3. A variety of cells seem to contain osmotically active materials which cannot readily penetrate the nuclear envelope. The chemical and physical characteristics of these substances are, at present, unknown.

Much remains to be learned concerning the pathways available to materials crossing the nuclear envelope. Various pathways have been suggested, an important likely pathway appears to be the nuclear annuli. Even less is known of the specific mechanisms of exchange. The work which has been done suggests that enzymatic processes may be involved.

Bibliography

Abelson, P. H., and W. R. Duryee, 1949: Biol. Bull. 96, 205.

Afzelius, B. A., 1955: Exper. Cell Res. 8, 147.

Allfrey, V. G., R. Meudt, J. W. Hopkins, and A. E. Mirsky, 1961: Proc. Nat. Acad. Sci. 47, 907.

Anderson, N. G., 1953: Exper. Cell Res. 5, 361.

Anderson, E., and H. W. Beams, 1956: J. Biophys. Biochem. Cytol. 2, Suppl. 4, 439.

— and K. M. Wilbur, 1951: J. Gen. Physiol. 34, 647.

— — 1952: J. Gen. Physiol. 35, 781.

Battin, W. T., 1959: Exper. Cell Res. 17, 59.

Baud, C. A., 1959: in: Problèmes d'Ultrastructures et de Fonctions nucléaires (Ed. J. André Thomas). Paris, Masson et Cie., 1.

Beck, L. V., and H. Shapiro, 1936: Proc. Soc. Exper. Biol. Med. 34, 170.

Becker, F. F., and H. Green, 1960: Exper. Cell Res. 19, 361.

Bennett, H. S., 1956: J. Biophys. Biochem. Cytol. 2, Suppl. 4, 99.

Brachet, J., 1955: Nature 175, 851.

— 1956: Biochim. et Biophys. Acta 19, 583.

— 1957: Biochemical Cytology. Academic Press, New York.

Buck, J. B., and R. D. Boche, 1938: Biol. Bull. 75, 344.

Byers, T. J., D. B. Platt, and L. Goldstein, 1963: J. Cell Biol. 19, 453.

Callan, H. G., 1949: Hereditas, Suppl. 35, 547.

— 1952: Symposia Soc. Exper. Biol. 6, 243.

Clark, A. M., 1943: Austral. J. Exper. Biol. a. Med. Sci. 21, 215.

Clark, W. H. Jr., 1960: J. Biophys. Biochem. Cytol. 7, 345.

Churney, L., 1942: Biol. Bull. 82, 52.

Duryee, W. R., 1940: Proc. 8th Amer. Sci. Congr. 3, 45.

Feldherr, C. M., 1962 a: J. Cell Biol. 14, 65.

— 1962 b: J. Cell Biol. 12, 139.

— 1963: Fed. Proc. 22, 179.

— Unpublished observations.

— and A. B. Feldherr, 1960: Nature 185, 250.

— and J. M. Marshall, 1962: J. Cell Biol. 12, 640.

FICA, A., and M. ERRERA, 1955: Arch. Inter. de Physiol. et de Biochim. **63**, 259.
FISCHER, H., and L. WAGNER, 1954: Naturwissenschaften **41**, 532.
GALL, J. G., 1964: Protoplasmatologia V/2, 4.
GAY, H., 1956: J. Biophys. Biochem. Cytol. **2**, Suppl. **4**, 407.
GOLDSTEIN, L., 1958: Exper. Cell Res. **15**, 635.
— and C. V. HARDING, 1950: Fed. Proc. **9**, 48.
— and W. PLAUT, 1955: Proc. Nat. Acad. Sci. **41**, 874.
HAMBURGER, H. V., 1904: Bd. III, Wiesbaden.
HARDING, C. V., and C. M. FELDHERR, 1959: J. Gen. Physiol. **42**, 1155.
HOLTER, H., 1959: Ann. N. Y. Acad. Sci. **78**, 524.
HOLTFRETER, V., 1954: Exper. Cell Res. **7**, 95.
HOLTZER, H., and S. HOLTZER, 1960: Compt. rend. trav. Lab. Carlsberg, Sér. Chim. **31**, 373.
HUNTER, A. S., and F. R. HUNTER, 1961: Exper. Cell Res. **22**, 609.
LOEWENSTEIN, W. R., and Y. KANNO, 1963: J. Cell Biol. **16**, 421.
MERRIAM, R. W., 1959 a: J. Biophys. Biochem. Cytol. **5**, 117.
— 1959 b: J. Biophys. Biochem. Cytol. **6**, 353.
MIRSKY, A. E., and S. OSAWA, 1961: in: "The Cell" (J. BRACHET and A. E. MIRSKY editors). Academic Press, New York, **2**, 677.
MONNÉ, L., 1935: Proc. Exper. Biol. Med. **32**, 1197.
MOORE, R. O., V. R. MUMAV, and M. D. SCHOENBERG, 1961: J. Ultrastruct. Res. **5**, 244.
MOSES, M. J., 1956: J. Biophys. Biochem. Cytol. **2**, Suppl. **4**, 397.
NACHMIUS, V. T., and J. M. MARSHALL, 1961: in: Biological Structure and Function. Proc. 1st IUB/IUBS Joint Symposium (T. W. GOODWIN and O. LINDBERG editors), New York, Academic Press.
NAORA, H., H. NAORA, M. IZAWA, V. G. ALLFREY, and A. E. MIRSKY, 1962: Proc. Nat. Acad. Sci. **48**, 853.
PAPPAS, G. D.,1956: J. Biophys. Biochem. Cytol. **2**, Suppl. **4**, 431.
POLLISTER, A. W., M. E. GETTNER, and R. WARD, 1954: Science **120**, 789.
SHINKE ,N., 1937: Cytologia, Fujii Jubilee Volumen 449.
STERN, H., and A. E. MIRSKY, 1953; J. Gen. Physiol. **37**, 177.
SWIFT, H., 1956: J. Biophys. Biochem. Cytol. **2**, Suppl. **4**, 415.
WATSON, M. L., 1955: J. Biophys. Biochem. Cytol. **1**, 257.
— 1959: J. Biophys. Biochem. Cytol. **6**, 147.
WHALEY, W. G., H. H. MOLLENHAUER, and V. H. LEECH, 1960: J. Biophys. Biochem. Cytol. **8**, 233.
WISCHNITZER, S., 1958: J. Ultrastruct. Res. **1**, 201.
— 1960: Inter. Rev. Cytol. **10**, 137.

Combined Nuclear Transplantation and Isotope Techniques for the Study of Nuclear Activities[1]

By

LESTER GOLDSTEIN

University of Pennsylvania

With 4 Figures

Contents

I. Introduction

Of the abundant techniques available to cell biologists, it is probable that almost all have been employed at one time or another to study some facet of the physiology of the cell nucleus. The rewards for these efforts—as can be seen by the other contributions to this number, as well as the reviews of BRACHET (1957), GOLDSTEIN (1963), MIRSKY and OSAWA (1961), and PRESCOTT (1960)—have been considerable. Although it is evident that the

---

[1] The work from our laboratory reported here has been supported by a research grant (RG-6774) from the United States Public Health Service.

usefulness of many of the techniques for studies of nuclear function is far from exhausted, this writer would like to draw some attention to a few valuable techniques that are at hand but appear to be underemployed. These latter techniques are based primarily on *nuclear transplantation* and *isotope methodology*—both of which have been extensively used, of course, but *not* in combination. This combination is very potent in the attack on a variety of problems, yet there are few reports on work in which it has been employed; the few reports will be referred to later.

The established and potential values of this combination of techniques are realistic only in experiments with large, free-living amoebae (*A. proteus*, for example) in interphase. These techniques are not known to the writer to have been applied together for investigations of nuclear behavior in other kinds of interphase cells. The greatest value in the employment of nuclear transplantation in combination with isotope techniques with amoebae lies precisely in the fact that the grafted nucleus is an *interphase* nucleus in, as we believe, a physiologically active stage. It is possible, of course, to transfer a nucleus (containing some kind of isotopic label) from one cell to another by fertilization of an egg, by conjugation between protozoa, or by some other sexual process—but in such cases the genetic material of the transferred nucleus is in a "condensed" state and is therefore, presumably, metabolically inert. Largely for this reason, these nuclei generally could not serve for the types of studies that will be described here. Transplanation of interphase nuclei and isotope technique used in combination have thus led to the strong confirmation or denial of evidence obtained by more traditional approaches to the same problems and also led to the discovery or investigation of phenomena that are inaccessible by other methods.

It is worth noting that certain conditions of the nucleus may be investigated through experiments with isolated nuclei *or* by the methods we are highlighting here. Since it is relatively easy to work with isolated nuclei and since quantitative results obtained with such nuclei may be more reliable, one might ask: Why devote time and energy on similar studies via nuclear transplantation experiments? The answer must be that the nuclei in the transplantation experiments are not exposed to environments other than cytoplasm and thus there is less concern that artifacts have been induced by experimental manipulation of the nuclei. Insufficient comparisons between the results of nuclear transplantation experiments and the results of isolated nuclei studies have been made to clarify the relative validity of the two approaches; one such projected study will be described here.

Some of the investigations that can be performed by our unique combination of techniques are: 1. a determination of whether certain biochemical substances are synthesized in the nucleus or the cytoplasm; 2. a study of the movement of some materials between nucleus and cytoplasm; 3. a study of the "non-movement," and thus a determination of the intracellular localization, of certain molecules; 4. the development of new and presumably more accurate, approaches for semi-quantitative or quantitative assays of the composition of the nucleus and of the cytoplasm; 5. investigation of the interspecies specificities of some kinds of macromolecules; and 6. a study of the effects of various intra- and extra-cellular factors on the above phenomena.

II. Methods

For an appreciation of our objectives in this article it is worth discussing, before considering the actual experimental designs, some of the relevant methodology.

The only nuclear transplantation operation referred to in this paper is a very slightly modified version of a method developed by COMMANDON and DE FONBRUNE (1939). It subsequently has been successfully employed in DANIELLI's laboratory (see e.g., DANIELLI et al. 1955), by GOLDSTEIN and PLAUT (1955), by IVERSON (1962), by CHADWICK (1961), and by BYERS et al. (1963). The fundamental procedure does not appear to differ significantly from one laboratory to another.

In all laboratories the DE FONBRUNE micromanipulator is used for the critical operations. This instrument seems to be the only commercially available one that will serve reasonably well for nuclear transplantations in amoebae because it combines two exclusive features that are virtually indispensible. One of these features is a remote, flexible connection between the hand-operated control lever and the microtool that engages the nucleus. This type of connection permits the microtools to be positioned relative to the hand operated controls in such a manner as to compensate, for the mechanically optically inverted image in a compound microscope and also eliminates the transmission of vibrations from the operator's hand to the microtools. The other feature is that 2 microtools can be mounted on the same manipulator head but only one of the microtools can be moved by the fine control mechanism, whereas both can be moved by the coarse control. The latter feature permits the use of only one micromanipulator and simplifies the construction of microtools.

The only version of the de Fonbrune instrument available in the United States (presumably because of trade restricting agreements between the American and French manufacturers not to infringe on one another's markets) is, in the opinion of this writer, one of the most poorly *constructed* micromanipulators available in this country: apparently the instrument manufactured in Paris is of a higher quality than the American one. Therefore, for those who are fortunate enough to have a choice or can prevail on European friends to assist, I recommend that they consider the purchase of the French model.

In preparation for the operation, three amoebae—the amoeba that will donate its nucleus, the amoeba that will receive the nucleus, and a third amoeba—are placed in a drop of culture medium approximately 10 times the volume of the amoebae on a coverslip that is inverted and placed over a shallow, oil-containing chamber (p. 129, DE FONBRUNE 1949) that has provision for the entry of tools on one side. Thus, the cells are in a hanging drop surrounded by oil. (The oil prevents evaporation of the aqueous drop and to some extent dampens vibrations of the microtools.) The third amoeba in the chamber is not involved in the transplantation but is sacrificed by rupturing prior to the actual operation. The liberated contents of the sacrificed amoeba serve to protect (in some unknown fashion) the remaining two amoebae from rupturing during subsequent maneuvers.

After the oil chamber is properly assembled, it is placed on the mechanical stage of a compound microscope and the cells are observed at a magnification of approximately 100–150×. Then the microtools—a glass probe approximately 2–5 μ at the tip, which is mounted on the manipulator to be operated by the pneumatic, fine controls, and a hook made of approximately 20–40 μ thick glass bent into a hook diameter of approximately 70–100 μ, which is mounted on the manipulator so as to be operated only by the coarse, mechanical controls—are carefully brought into the operating chamber via its open side. After the third amoeba is sacrificed, the remaining two amoebae are oriented appropriately with respect to one another and surrounded by the hook (which thus immobilizes them to some extent). The fine probe is then inserted into the donor and the nucleus pushed with care from one cell to another *without exposing the nucleus to the external medium.* Upon withdrawal of the probe from the cells, all cell surfaces "heal" more or less instantaneously. Should the experiment require that the cell receiving the nucleus be without its original nucleus, that nucleus is pushed out of the cell (and into the surrounding medium) before the new nucleus is inserted. After all operations are complete, the cover-slip is removed and the cell (or cells) is recovered with a braking pipet (HOLTER 1943) and placed in a standard growth chamber.

Since—in most experiments—the only difference between donor and recipient cells is that one is labeled with an isotope and the other is not, the cells are made visibly distinguishable for microscopic purposes by feeding ciliate food organisms to one amoeba just before the operation and starving the other for approximately one day. The fed amoeba then can be readily identified by the presence of microscopically conspicuous food vacuoles.

The detection of isotopes can be carried out by standard methods and requires no discussion here. Radioisotopes are detectable by Geiger counting, scintillation counting, and by autoradiography—with the latter method predominating, since we are interested in intracellular distribution of label. Experiments with stable, heavy isotopes are also possible and detection of these will be described in the relevant section below.

III. Experimental Potentials

Let us now turn to some of the proven and some of the anticipated potentialities of experiments employing nuclear transplantation and isotope techniques in combination. The reader will shortly realize that these techniques have been exploited only to a rather limited extent and that a little additional imagination can readily find expression in new experimental designs. Because of the limited applications of these methods the illustrations given here will vary in clarity—some perhaps will be vivid, some sketchy, and some purely imaginary.

A. Movement of Macromolecules

This combination of techniques was employed first to observe the movement of molecules between nucleus and cytoplasm (GOLDSTEIN and PLAUT 1955). This perhaps remains the most rewarding use to which these techniques have thus far been put. Probably the only improvement on observations of this type will come from a direct visualization of the behavior of specific molecules in *living cells*—a procedure that does not appear available for imminent practical laboratory use. The following studies on the movement of RNA and of protein will demonstrate how much more direct than others are our techniques.

1. R N A

Many investigators have studied what they believed to be the movement of RNA between nucleus and cytoplasm and the majority observed this "movement" by following the intracellular course of label following a brief exposure of the cells to a radioactive RNA precursor. After such a brief exposure—at which time all the incorporated label appears to be restricted to nuclear RNA—the cells are placed in a medium containing relatively high concentrations of non-radioactive RNA precursors (see e. g., WOODS and TAYLOR 1959, ZALOKAR 1959, GOLDSTEIN and MICOU 1959, and HARRIS 1959). Continuous observations, usually by autoradiography of fixed cells, of samples taken from the labeled cell population showed that after transfer to the non-radioactive medium (chaser) the amount of label in the nucleus diminished with time while an almost reciprocal increase was observed in the amount of cRNA[2] label. These data have usually been taken to indicate that RNA moves from nucleus to cytoplasm and therefore it is assumed that nRNA is a precursor of cRNA.

HARRIS (1959) has emphasized that these interpretations do not give serious consideration to the very reasonable possibility that at the time of change from a radioactive to a non-radioactive medium there exists within the cell a pool of radioactive RNA intermediates that is not diluted by the presence of excess non-radioactive precursors in the new environment. If HARRIS's suggestion is true, it is reasonable to suspect the nRNA and cRNA have different turnover rates, the former being more rapid, and that it is more or less coincidental that as the nRNA label is decreasing the cRNA synthesizing machine (in the cytoplasm) begins to incorporate the residual, intracellular labeled intermediates into high molecular weight cRNA.

The above ambiguities about the behavior of RNA can be clarified to some extent by the more direct approach afforded by experiments involving the transplantation of labeled nuclei. This has been done (GOLDSTEIN and

---

[2] cRNA = cytoplasmic RNA; nRNA = nuclear RNA.

PLAUT 1955) by transplanting a nucleus containing P^{32}-labeled RNA into an unlabeled cell and following the subsequent fate of the label by autoradiography. It is evident in such experiments—and contrary to those described above—that interpretations are not qualified by complexities due to the presence of a pool of labeled precursors in the cytoplasm, since *all* the radioactivity must come from the nucleus. (If any radioactive cytoplasm was also transplanted from the P^{32}-labeled cell, it contributed so small a proportion of the total transferred radioactivity as to be undetectable by the assay methods used.) Analysis of the autoradiographs showed that the radioactivity appeared in the cytoplasm shortly after the implantation of the P^{32}-labeled nucleus and that the proportion of radioactivity in the cytoplasm apparently rose continuously all through interphase—although it is difficult to establish with confidence that it was a continuous process. Tests of ribonuclease digestibility revealed that the label, which had initially been in RNA in the nucleus, was restricted to RNA in the cytoplasm. Finally, it was observed that very little, if any, radioactivity ever appeared in the recipient cell nucleus—suggesting that there is some kind of specificity imparted to the migrating label.

The results of the nuclear transfer experiments support the conclusion that RNA can pass from nucleus to cytoplasm—as has been suggested by a variety of other experiments—and that probably little, if any, passage of RNA proceeds from cytoplasm to nucleus. These conclusions are subject, however, to one important qualification: It may be that the labeled-RNA-containing nucleus also contained a substantial pool of P^{32}-labeled precursors (nucleotides), that this pool of label was incorporated into cRNA by synthesis in the cytoplasm, and that there thus may be virtually no passage of high molecular weight RNA from nucleus to cytoplasm. The question of whether there are sufficient P^{32}-labeled nucleotides in the transferred nucleus to account for the amount of cRNA label will be considered below in the section on the site of cytoplasmic RNA synthesis (Section III C 3).

2. Protein

The movement of RNA from nucleus to cytoplasm is in accord with theoretical expectations that are at least 10 years old (see e. g., DOUNCE 1953 and RICH and WATSON 1954). Since that time there has been widespread suspicion (almost proven now) that RNA is the carrier of information from gene to protein, which would lead to the expectation that RNA should be produced in the nucleus and then be passed to the cytoplasm. But what can we expect of the movement of *protein* between nucleus and cytoplasm? No speculations of any order have provided clues to the answer to this question. It may be that protein moves only from nucleus to cytoplasm, only from cytoplasm to nucleus, in both directions, or moves not at all between the two compartments; no one, apparently, has even ventured to theorize seriously on this matter. Studies of the movement of protein between nucleus and cytoplasm might provide insight into nucleocytoplasmic interactions, if the movement were to prove to be in some way unique. As it has developed, very unexpected and unique discoveries have been made and it is probable that this could not be done with techniques other than those that were actually employed.

The first experiment concerned with the movement of protein would obviously be designed to determine what would happen if a nucleus from a cell labeled by incorporation of a radioactive amino acid into protein is grafted into an unlabeled cell. When this experiment was performed (Goldstein 1958) it was immediately apparent that the labeled material (containing S^{35}-methionine in this case) was localized almost completely in the donor nucleus *and* the host nucleus shortly after the operation; very little radioactivity could be detected in the cytoplasm—even after long incubations. These observations now have been repeatedly confirmed (Byers et al. 1963). Since autoradiographs of the labeled donor cells show approximately uniform labeling over nucleus and cytoplasm and since brief exposure to radioactive amino acids does not reveal any significant difference between the rates of incorporation of amino acids into nucleus and cytoplasm, it is highly unlikely that the aforementioned observation of protein movement could have been made by other methods currently at the disposal of cell biologists. (Indeed, these observations would be unlikely even if there had been differential labeling of nucleus and cytoplasm, since the observations appear to be completely dependent on having an initial state when one nucleus is labeled and the other in the same cell is not—a condition impossible to achieve without micromanipulation.) This migratory behavior of protein can be inferred, however, from other kinds of data (Prescott 1963) in the light of the nuclear transplantation experiments and has been confirmed, after a fashion, by Kroeger et al. 1963.

The results of the above experiments, along with a series of other observations (Byers et al. 1963), strongly urge the view that there exists a class of cellular proteins—designated *cytonucleoproteins* by Byers et al.—that is in constant, *non-random* migration between nucleus and cytoplasm and that are present in the nucleus at a concentration at least 20 and perhaps 240 more times than that in the cytoplasm. Many of these experiments have almost conclusively established that the migrating material is in protein and not in the form of amino acids or small peptides. Different kinds of experiments have also been concerned with various other aspects of cytonucleoprotein behavior, such as metabolic stability, fate at mitosis, etc.

Although we are still ignorant about the *function* of the cytonucleoproteins, we are now aware of their existence. This awareness probably would still be far in the future had it not been for experiments combining nuclear transplantation and autoradiography. These have been rewarding experiments because it is apparent from the behavior already noted that these proteins probably play an intriguing role in the physiology of the cell.

B. Localization of Macromolecules

The intracellular localization of macromolecules is the object of almost countless studies by investigators of various disciplines—but particularly by those who are called cytochemists. A great deal has been learned from their efforts and many of their techniques are those of choice for detecting particular types of molecules. But, as we have just seen, some kinds of macromolecules can not be identified by methods employed in most cytochemical laboratories. The cytonucleoproteins were characterized largely on the basis of their migrating behavior; it is possible to recognize still other classes of molecules on the basis of other kinds of behavior following nuclear transplantation between labeled and unlabeled cells.

1. Discovery of Non-migrating Nuclear Proteins

We mentioned earlier that when a nucleus from a cell that has been labeled with a radioactive amino acid is grafted into an unlabeled cell, the radioactivity comes to be localized almost exclusively in the protein of both the grafted and host cell nuclei. The data, furthermore, strongly support the conclusion that this material—cytonucleoprotein—is in constant migration between nucleus and cytoplasm, being present at any instant in much higher concentration in the nucleus. Under such circumstances one would expect that at some equilibrium stage the two nuclei would be more or less equally labeled in the transplantation experiments. Analysis of the experimental material, however, shows that at equilibrium (which is reached about 5 hours after the implantation of the labeled nucleus) the two nuclei are *not* equally labeled; indeed, the host nucleus is found to contain only about 30% of the total activity of the two nuclei (BYERS et al. 1963).

Interpretation of this inequality leads to the expectation that a *non-migrating* class of proteins is present in the nucleus. This conclusion is supported by a series of experiments involving the serial transplantation through several cells of the nuclei from the first experimental cell. Thus, when the original grafted nucleus is *re*-transplanted to a fresh unlabeled cell (see Fig. 1 to follow this experiment) the new host nucleus acquires not more than 20% of the total radioactivity of the two nuclei in that cell. On the other hand if—after the equilibrium distribution is reached in the first host cell—the host cell nucleus is transplanted to a fresh unlabeled cell, the two nuclei in this latter cell become approximately equally labeled after the equilibrium distribution of radioactivity is once again attained. These observations, therefore, appear to be most consistent with the view that the original radioactive nucleus contains a class of proteins whose label does not pass through an interphase nuclear membrane. Thus, upon serial transplantation this nucleus loses more and more of the cytonucleoprotein label but not the non-migrating protein label and we observe that it retains a higher and higher proportion of the total nuclear radioactivity in a binucleate cell. The original unlabeled nucleus (of the first host cell), which acquires only cytonucleoprotein label, is able to share its radioactivity almost equally with subsequent host cell nuclei after serial transplantation.

We thus have discovered a previously unrecognized class of proteins that is *localized* almost exclusively in the nucleus; there is no reason to believe that these proteins have ever been consciously studied before. There are two reasons to suspect that the non-migrating proteins are not in any way permanent components of the chromosomes. First, since the labeling pattern with radioactive tryptophan and arginine does not differ from that when other labeled amino acids are used, we suspect that histones are not involved. Second, we have noted that at mitosis the label of the non-migrating nuclear proteins (as well as that of the cytonucleoproteins)

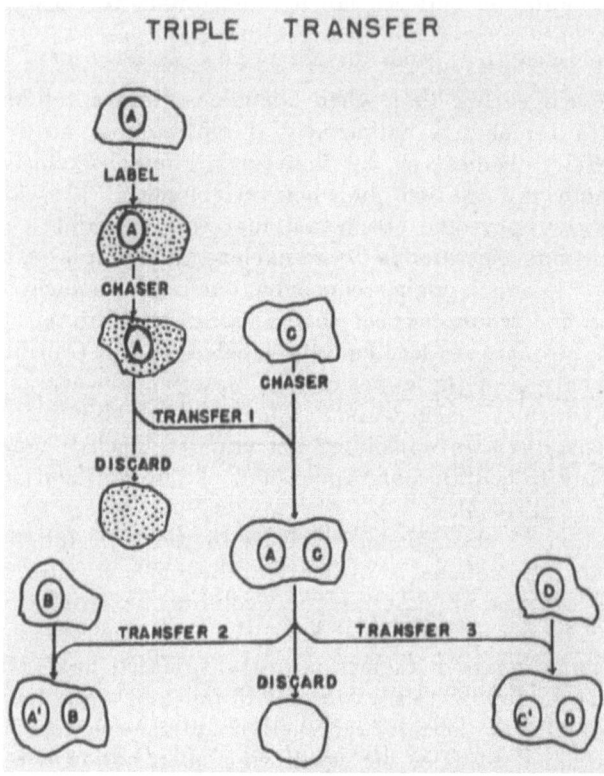

Fig. 1. Diagram of experiment described in the text. Nucleus from uniformly labeled cell (A) is grafted, afte period in chaser, into unlabeled cell (C), which had been preincubated in chaser. After time for equilibrium to be established, nucleus A is grafted into unlabeled cell B and nucleus C is grafted into unlabeled cell D. After equilibrium is again established the radioactivity in nucleus A' is to the radioactivity in nucleus B in the proportion of ca. 80:20; the radioactivity is distributed between nuclei C' and D in the proportion of ca. 50:50.

appears to be uniformly distributed throughout the *entire* cell and is in no detectable fashion associated with the condensed chromosomes. The only known substances of relatively high nuclear concentration that might be identical with the non-migrating proteins are the few enzymes associated with the synthesis of certain nucleotides (Hogeboom and Schneider 1952) but it is doubtful that these latter proteins are present in concentrations sufficient to account for the pattern of labeling that is observed.

Given the ability to tag other macromolecules in similar ways, it should be possible to determine unique localizations of other substances of potentially great interest.

C. Site of Synthesis of Cellular Components

Perhaps the greatest value of these techniques to cell research is the potential for determining the site of synthesis of a variety of cellular components. Although it is possible that the site of synthesis of *some* of these components may be more favorably investigated by other contemporary methods, it is doubtful that any techniques—other than the combination of nuclear transplantation and isotope techniques, where applicable—can give answers to the same questions as directly. Conversely, some of these less direct techniques can give answers that are more precise in a number of respects. For example, our methods can provide (see below) the most direct evidence that at least some cRNA is synthesized in the nucleus whereas PERRY (1962), e. g., is able by a combination of other techniques to demonstrate which class of RNA molecules probably comes from which part of the nucleus. As will be suggested below, it is possible that our methods also may provide, in time, more direct evidence for the same relationships being illuminated by PERRY and others.

1. Site of Cytonucleoprotein and Non-migrating Nuclear Protein Synthesis

The presence of the cytonucleoproteins and the non-migrating nuclear proteins in the nucleus at much higher concentration than in the cytoplasm can be taken to signify, as has occasionally been the case for other substances of localized high concentration, that these proteins are synthesized where they largely localized. However, this obviously can not be considered very strong evidence for such a conclusion; the same logic could be invoked, for example, to prove the unlikely conclusion that cellulose is synthesized in the non-living cell walls of plant tissue. For a variety of theoretical reasons we wished to determine, with greater certainty, whether the aforementioned proteins are synthesized in the nucleus.

The site of cytonucleoprotein and non-migrating protein synthesis has been investigated intensively (BYERS et al. 1963). The experiments in this part of the investigation were based largely on the fact that enucleate amoebae can incorporate radioactive amino acids into TCA (trichloroacetic acid) insoluble material and this incorporation is presumed to be a reflection of protein synthesis. Consequently, if enucleate amoebae synthesize cytonucleoproteins and non-migrating proteins, this should be demonstrable by our techniques.

Unlabeled amoebae were enucleated and incubated for several hours in a medium containing a radioactive amino acid—during which time an appreciable amount of radioactivity was incorporated into TCA-insoluble material. The enucleate cells then were placed in non-radioactive medium for periods long enough to virtually deplete any residual intracellular pool of unincorporated radioactive amino acids. Into each labeled enucleate cell was implanted an unlabeled nucleus and the renucleated amoeba was incubated for at least another 5 hours—long enough for equilibrium of labeled cytonucleoprotein movement between nucleus and cytoplasm to be

established. This incubation was followed by the retransplantation of the
nucleus to an unlabeled, unenucleated cell and, following another incubation
of at least 5 hours, the final binucleate cell was fixed and subjected to
autoradiography. These operations are diagrammed in Fig. 2.

ENUCLEATION AND DOUBLE TRANSFER

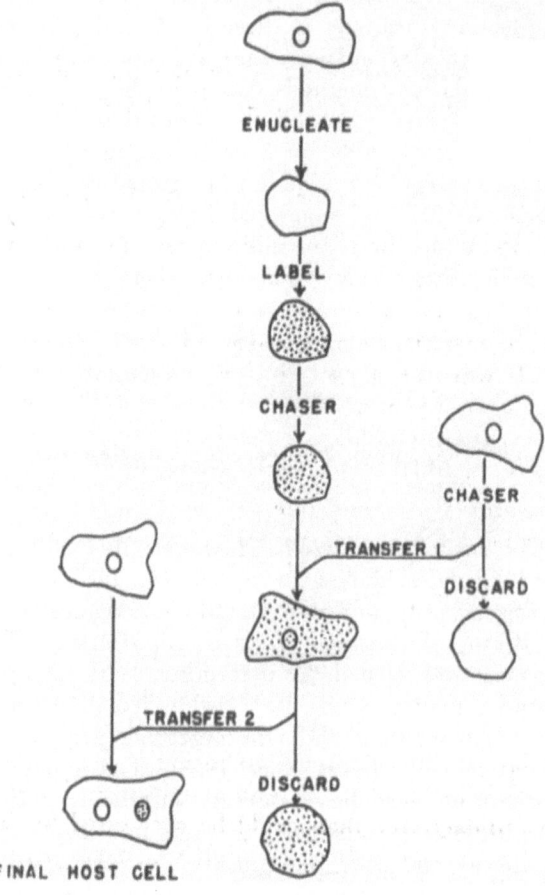

Fig. 2. Diagram representing experiment to determine localization of cytonucleoprotein synthesis. Following
removal of its nucleus, cell is incubated in leucine-H³ and then in chaser. At the next step an unlabeled nucleus (preincubated in chaser) is grafted into the enucleate cell (Transfer 1). After further incubation the nucleus is
transferred (Transfer 2) to an unlabeled cell (Final Host Cell) that is incubated in unlabeled medium before fixation.

Examination of the autoradiographs of the final cells in the series reveals
that the radioactivity is localized almost completely in the two nuclei—as
is the case when a nucleus from a cell uniformly labeled with radioactive
amino acids is grafted into an unlabeled cell. Moreover, assay of the
distribution of radioactivity between the two nuclei shows that the nucleus
of the final recipient cell acquires only about 30% of the total radioactivity
of the two nuclei. Thus, we see that the nucleus grafted into the radioactive enucleate cell has apparently acquired labeled cytonucleoproteins

and labeled non-migratory proteins, and this is substantial evidence that both classes of proteins are synthesized in the cytoplasm. However, there is as yet no unequivocal proof that the nucleus initially grafted into the radioactive enucleate cell did not synthesize these labeled proteins from residual intracellular, unincorporated radioactive precursors present in the cytoplasm, but this appears unlikely in the light of a variety of experimental data dealing with the possibility of such a mechanism (BYERS et al. 1963).

It seems *possible,* in view of the results from the above experiments, that no proteins are synthesized directly in the nucleus—although this suggestion is, of course, not strongly substantiated by the information so far in hand.

2. Site of Nuclear RNA Synthesis

Approximately 10 years ago an interest in the nature of genetic "messengers" and their involvement in nucleocytoplasmic relations led to the first experiments in which nuclear transplantation and isotope techniques were used in combination. Because of the then relatively new suspicion that RNA was the agent for the transmission of information from gene to cytoplasm (DOUNCE 1953 and RICH and WATSON 1954), RNA commanded primary attention. Our original studies and, of course, much recent research have justified those early speculations.

If the genes are directly responsible for the synthesis of "messengers" and if these are made of RNA, then we should expect to find that at least the RNA present in the nucleus at any moment is synthesized there. This view can readily be confirmed by the following experiment (which has been described in some detail in Sect. III A 1). A nucleus with radioactive RNA is grafted into an unlabeled cell and incubated until fixed several hours later. Autoradiographic analysis reveals that little, if any, of the radioactivity that passes out of the transplanted nucleus goes to the recipient cell nucleus. If the reciprocal operation is performed (the transplantation of an unlabeled nucleus into an enucleate cell with radioactive RNA), little more radioactivity than could be accounted for by the availability of labeled precursors resulting from cRNA turnover is found in the implanted nucleus. Both of these observations strongly suggest that RNA can not pass from cytoplasm to nucleus as an intact molecule and we conclude, therefore, that nRNA must be synthesized in the nucleus.

IVERSON (1962), who used techniques similar to ours, has recently reported the passage of labeled RNA from cytoplasm to an implanted unlabeled nucleus. The experiments are not described in sufficient detail to make a critical judgment but certain technical steps that are described are open to some question. We, too, have observed the appearance of radioactivity in the recipient cell nucleus on occasion but, although we have as yet no explanation for this type of labeling, we have found the host cell nuclear radioactivity resistant to ribonuclease digestion. Further experiments are needed to clarify these puzzling observations.

3. Site of Cytoplasmic RNA Synthesis

On the basis of the earliest studies of the kinetics of incorporation of RNA precursors into nucleus and cytoplasm, as well as other kinds of studies, there was seldom serious doubt that nRNA was synthesized in the nucleus. What remained—and in a few quarters still is—in doubt was the site of cRNA synthesis: Is cRNA synthesized in the nucleus as we would expect a genetic "messenger" to be or is it synthesized in the cytoplasm?

The answer comes in part from the same experiments as described above on the site of synthesis of nRNA. Although the radioactivity of the RNA of the grafted nucleus does not pass to the recipient cell nucleus, it does pass into the cytoplasm. This cytoplasmic radioactivity is found to be still in RNA as indicated by its susceptibility to removal by ribonuclease. Therefore, since the label in nRNA later appears in cRNA and since the label has some specificity which inhibits its passage to another nucleus, it may be concluded that cRNA is probably synthesized in the nucleus. .

An alternative interpretation of the same data, however, is that the grafted nucleus contained not only radioactive high molecular weight RNA but also radioactive RNA *precursors* (as an acid-soluble pool undetectable by our autoradiographic methods), that would be preferentially incorporated into cRNA by synthetic mechanisms outside the nucleus. This possibility appears unlikely in view of repeated observations (Woods and Taylor 1959, Zalokar 1959, Goldstein and Micou 1959, and Harris 1959) that RNA precursors are incorporated more rapidly by the nucleus than the cytoplasm, a circumstance that contrasts with the observation that the recipient cell nucleus acquires little of the radioactivity brought into the cell by a grafted nucleus. Although improbable, this alternative interpretation is still not excluded, however, since it is possible that intranuclear precursors are more directly specific for the cytoplasmic synthesis of RNA than are precursors added to the environment of cells by the investigator. Consequently, further experiments are needed to establish the site of cRNA synthesis.

One seemingly direct approach to this problem is to determine whether enucleate cells can incorporate precursors into RNA. An answer in the affirmative should be relatively unambiguous evidence that cRNA is synthesized in the cytoplasm. However, data from various sources show that, except for instances where DNA-containing structures may be present in the cytoplasm, probably no enucleate cell is capable of incorporating radioactive RNA precursors (see e. g., Prescott 1961 and Goldstein et al. 1960). Unfortunately such evidence is not strong support for the conclusion that no RNA synthesis occurs in the cytoplasm. The inability to incorporate precursors can be explained just as well by the dependence of some synthetic machinery in the cytoplasm on an unstable factor furnished by the nucleus, which thus would be involved only indirectly in cRNA synthesis.

We can improve upon enucleation experiments in our efforts to clarify the nature of the nuclear contribution to cRNA and reference to Fig. 3 should help in following the details of the experiment. (For greater details of the experiments see GOLDSTEIN 1963 a). Recall that under the usual culture conditions when a radioactive RNA precursor is available to whole cells, the nucleus becomes labeled earlier than does the cytoplasm. Thus, if cRNA is not synthesized in the nucleus, we would expect that the cytoplasmic machine was slower than the nuclear machine. Therefore, upon incubation of amoebae for one hour or so in the presence of radioactive

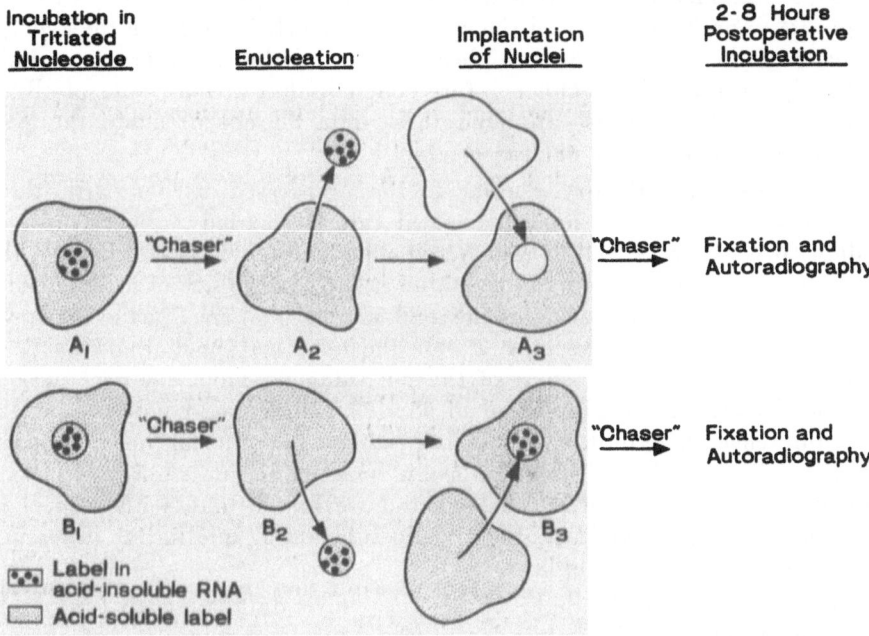

Fig. 3. Diagram of an experiment described in the text. Amoebae A_1 and B_1 were incubated for 1-4 hours in a medium containing tritiated nucleoside. Upon removal to chaser they were enucleated (A_2 and B_2). Then into A_3 was grafted a nucleus from an unlabeled amoeba and into B_3 was grafted a nucleus from an amoeba kept in the same medium with A_1 and B_1. After 2-8 hours in chaser A_3 and B_3 were fixed together on the same slide.

RNA precursors (A_1 and B_1 in Fig. 3) the nucleus will have incorporated label into high molecular weight RNA and the cytoplasm would have not—but the cytoplasm would presumably contain a supply of all precursors and intermediates (all acid-soluble) that go to form macromolecular RNA. At this stage one can remove the nucleus with radioactive RNA (A_2) and replace it with an unlabeled nucleus (A_3). Thus, RNA synthesis in the cytoplasm, if it occurs, can proceed with a nucleus essentially present continuously but not furnishing any radioactive RNA to any part of the cell. For an ideal experiment, one would expect to find that the newly grafted nucleus does not become radioactive at any time thereafter, but unfortunately the ideal is not attained and one can never be certain, therefore, that any subsequent cytoplasmic radioactivity does not somehow come from the nucleus after all. To circumvent *this* difficulty one can perform the control shown in the B_1-B_2-B_3 series in Fig. 3. Then we can reason that, if cRNA synthesis occurs outside the nucleus, the amount of cytoplasmic radioactivity in RNA in A_3 and B_3 should be the same—there being no contribution from the nucleus. If there is would-be cRNA inside the nucleus and consequently an expected nuclear contribution, the cytoplasm of cell B_3 should be substantially more radioactive than the cytoplasm of A_3—even though the nucleus of A_3 were to become radioactive (which it does). Although the data prove to be cruder than the ideal possible to obtain by our methods and therefore tend to minimize the differences between the cytoplasms of the two cell types, 24 experiments showed the cytoplasm of cell B_3 to be on the average twice as radioactive as the cytoplasm of A_3 and the difference is statistically significant. One might therefore conclude that *at least* 50% of cRNA comes from the nucleus—although these data do not preclude the possibility that 100% is of nuclear origin.

One obvious difficulty, similar to that expressed earlier, with these experiments is that the nucleus grafted into B_3 carried radioactive acid-soluble precursors as well as radioactive macromolecular RNA and that the greater cytoplasmic labeling in B_3 is due to the availability of a larger total precursor pool. Examination of the acid-soluble pool size, however, shows that the nucleus contains only 2% of the total cell pool (and this is in proportion to the nuclear share of the cell volume). Thus, the pool sizes of A_3 and B_3 are in the proportion of 98:100, which could not account for the difference in cytoplasmic labeling—unless the nuclear pool is *qualitatively* different from the cytoplasmic pool. This possibility has been investigated (see Section D 1 below) and we found that, with respect to the nucleotide pool at least, there is no substantial qualitative difference between nucleus and cytoplasm.

We conclude, in the light of all the above data, that acid-insoluble (macromolecular) nRNA (which is synthesized in the nucleus) is a precursor of acid-insoluble cRNA. Whether *all* cRNA derives from the nucleus is undetermined by these experiments but there is the suggestion that at least 50% is.

4. Site of Ribosome Synthesis

Many studies on the site of synthesis of RNA, such as those cited above, failed to consider that there are (at least) 3 general classes of RNA: ribosomal, "messenger," and transfer or amino acid-acceptor. Recently, however, we have been able to envision methods for discriminating between the various RNAs in our experiments. Taking a lead from Brenner et al. (1961), it now seems feasible—through the use of stable isotopes—to determine whether ribosomal RNA, e. g., is synthesized in the nucleus.

Rather than investigate the origin of ribosomal RNA, however, it is experimentally more feasible to inquire first into the origin of the complete ribosome, that is the RNA and protein components as a unit. The experiment, in essence, should be quite simple: amoebae will be grown on a food source that is heavily labeled with the heavy isotopes N^{15} and C^{13} and the radioactive isotope P^{32}. (After growth on this food the nucleoproteins—as well as other substances, of course—should be "heavy" and radioactive.) A nucleus from such a triply labeled cell therefore should have "heavy" and radioactive ribosomes, if ribosomes are synthesized in the nucleus, and when transferred to an unlabeled cell should pass triply labeled ribosomes to the cytoplasm. It only remains to determine whether "heavy" and radioactive ribosomes are then found in the cytoplasm; and this looms as the major technical difficulty of the experimental procedure.

Fortunately, BRENNER et al. (1961) have developed a splendid technique for separating heavy (N^{15}, C^{13}) ribosomes from light (N^{14}, C^{12}) ribosomes. Since each type of ribosome has a different density, they can be clearly separated from one another by prolonged centrifugation in a continuous density gradient of cesium chloride until each type reaches its buoyant density. Thus, in our nuclear transfer experiments it presumably would be necessary to determine whether or not ribosomes extracted from the cytoplasm had a buoyant density equivalent to those from cells grown on N^{14} and C^{12}.

In practice, the above experiment is almost impossible to perform, since it would be an overwhelming task to perform enough nuclear transfers to subsequently provide sufficient "heavy" ribosomes at a particular density position for analysis by ordinary spectrophotometric means. It is because of this difficulty that we propose to label with P^{32} also and trust that relatively few triply labeled nuclei will be needed to furnish enough radioactive ribosomes for assay by standard Geiger counter techniques. The actual procedure should be as follows: After 20–50 triply labeled nuclei are grafted into unlabeled cells and incubated for 12–24 hours, the cytoplasm will be extracted for ribosomal material; the experimental ribosomes then will be mixed with large amounts (in terms of U. V. absorbance) of previously prepared $C^{12}N^{14}$ ribosomes and $C^{13}N^{15}$ ribosomes and centrifuged in cesium chloride until equilibrium buoyant densities are reached; and, finally, the various layers of the mixture in the centrifuge tube will be assayed for U. V. absorbance and radioactivity.

We expect to find two major layers of U. V. absorbing material: one layer corresponding to light ribosomes and the other heavy ribosomes. The distribution of radioactivity can not be predicted but the following observations, and interpretations, seem possible: a) The radioactivity may be restricted completely to the layer of "heavy" ribosomes. This would imply that the *complete* ribosome is synthesized in the nucleus; b) The radioactivity may be restricted to a band somewhere between the "heavy" and light ribosome layers. This would suggest that only *ribosomal RNA* is synthesized in the nucleus, since such a buoyant density would imply that the RNA (distinguishable by P^{32}) is "heavy" and the protein is light (and therefore of cytoplasmic origin); c) The radioactivity may be found distributed through the centrifuge tube or in other discrete layers. The interpretations of these latter observations would be more uncertain and thus not worth elaborating here. The latter results would suggest, however, that ribosomes (or ribosomal RNA) are not synthesized in the nucleus.

Should the site of ribosome synthesis be determined from these experiments, we expect that the investigations of the origin of the other types of RNA will have greater significance and be more meaningful than heretofore.

D. Analysis of the Composition of the Nucleus and of the Cytoplasm

Quantitative analyses of the composition of nuclei are almost always performed by one of two general procedures. The cytologist's procedure is that of *in situ* spectrophotometry (Walker and Richards 1959) and has been usefully applied to study the concentrations of DNA, RNA, and protein in different parts of cells. The determination of protein is restricted essentially to the total protein population of the area being measured or to the presence of particular kinds of dye-binding side groups in the composite mixture of proteins. Small molecules of all sorts—particularly those soluble in the common fixatives—are difficultly or not at all preserved and hence are usually lost to the analysis. There are thus many kinds of molecules that are essentially unmeasurable by microspectrophotometric means. The other general procedure is the so-called biochemical one and is dependent on the mass isolation of nuclei from cells followed by the application of analytical techniques readily available to biochemists (Allfrey 1959). The nuclear isolation procedures are, however, laden with artifact-producing hazards (Allfrey 1959 and Roodyn 1959) and even the so-called non-aqueous methods have not been shown to be free of error, although the arguments offered in their favor are compelling. Moreover, heterogeneities in the cell population are undetected. Quantitative analyses of the cytoplasmic contents are also not completely dependable since it is necessary to perform assays first on whole cells (no necessary difficulty here) and then presumably subtract the values obtained for the isolated nuclei. If one could be certain that the nuclei are free of artifact, if the cell population were truly homogeneous, and if the yield of nuclei is determined with accuracy, one might then feel confident that the determinations for the cytoplasm were valid.

Faced with these difficulties and the requirements of our investigations, we have begun to develop procedures for more precise analysis on a few cells of some nuclear and some cytoplasmic components. (An additional harvest of this effort may be the possibility of determining the reliability of nuclear isolation methods.) The method we are developing is a combination of: a) a tracer technique—to amplify the chemical "signal": and b) nuclear transplantation—to figuratively isolate the nucleus. The tracer technique involves detection of a specifically labeled radioactive substance present in quantities too small to be analyzed chemically; "isolation" by nuclear transplantation involves isolating the nucleus from its original cytoplasmic environment into a new cytoplasmic environment. in which the assay is more feasible, *without subjecting the nucleus to the hazards of the standard isolation procedures.*

1. Measurement of the Nuclear and Cytoplasmic Nucleotide Pools

The actual procedure is illustrated best perhaps by our efforts to determine the composition of the nuclear and cytoplasmic nucleotide pools (GOLDSTEIN 1963 a). The motivation for doing these experiments is cited in Section III C 3. Amoebae are fed heavily labeled P^{32}-*Tetrahymena* and, after two or three cell growth cycles on this food, are placed in a non-nutrient

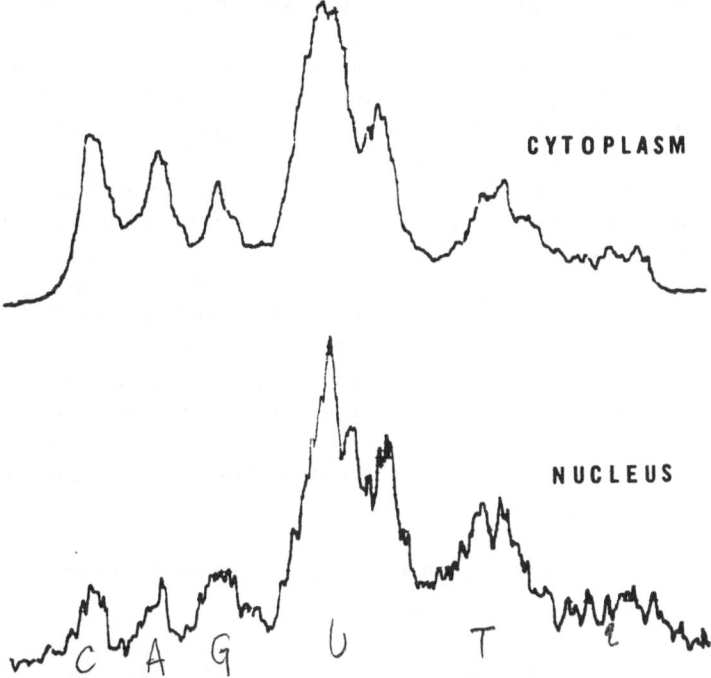

Fig. 4. P^{32} activity "profiles" of electrophoretograms of the acid-soluble material of nucleus and cytoplasm, subjected to electrophoresis for separation of nucleotides. The "profile" for the cytoplasm represents a scan at higher radioactivity scale and a slower time constant than the scan for the nuclear electrophoretogram. Beginning at the left, the order of the peaks corresponds to the following order of added carrier material: origin, cytidylic acid, adenylic acid, guanylic acid, uridylic acid, thymidylic acid, and the remainder is unidentified.

medium for approximately 24 hours. At that time the P^{32}-labeled nucleotide pool presumably would have attained a stable concentration (but this has not yet been determined and is believed to be not particularly important to the analysis). These cells are used thereafter as a source of nuclei and cytoplasm: the nuclei are transplanted into unlabeled cells that are dried on a planchet within 2 minutes of the operation and therefore are considered to be nuclei (containing P^{32}-labeled nucleotides) "isolated" in a sac of unlabeled cytoplasm; the enucleate P^{32}-labeled cells are dried on another planchet within 2 minutes of the operation and serve as the sample of isolated cytoplasm. The dried preparations, containing 30–100 labeled nuclei on one planchet and an equal number of labeled enucleate cells on the other, are extracted with 45% acetic acid and the extracts (mixed with carrier unlabeled nucleotides) are subjected to paper electrophoresis for separation of monophosphate nucleotides. (Analyses show that approximately 2% of the acid-soluble P^{32} is in the nucleus and this corresponds to the proportion of the cell volume occupied by the nucleus.) The electrophoretically separated monophosphate nucleotide spots are localized by the U. V. absorption of the carrier nucleotides and the radioactivity distribution determined by a paper strip scanning, gas-flow, Geiger counter (Model 880 Autoscanner, Vanguard Instrument Co.). The results of one experiment are shown in Fig. 4. Within the limits of the processing and assay techniques, the "profiles" for nucleus and for cytoplasm are considered to be very similar.

While this procedure provides only a relative quantitative measure of the nucleotide pool (an absolute determination probably can be made by isotope dilution techniques), it nevertheless appears to furnish a fairly precise comparison of these particular components in the two cell compartments. It is evident that few, if any, of the numerous proposed objections to standard nuclear isolation procedures can apply to the methods described above and one may confidently conclude, for these experiments, that the nuclear membrane probably is no real barrier to the free passage of small nucleotides. Given intensive enough labeling and as ready a specific identification of the compounds, it is probable that similar assays for the distribution of other substances in nuclei and cytoplasm can be made. It will be of interest also to compare the nuclear nucleotide pool composition by our methods with the composition observed in nuclei isolated by aqueous and non-aqueous procedures and, thus, determine the reliability of the various procedures.

E. Interspecies "Specificities" of Macromolecules

There are variety of methods for studying the specificities of closely related macromolecules; singly or in combination these methods include: immunochemistry, sequence determinations of sub-units, enzymatic reactivities, electrophoretic and chromatographic properties, etc. Since these are all *in vitro* methods, the possibility of artifact is difficult to exclude and the possibility of missing subtle sub-molecular differences may be low. We suspect that the techniques being promoted here may provide insight, at a very sensitive *in vivo* physiological level, into the degree of similarity between related molecules from different species. The method we propose may be considered the macromolecular analog of specificity determinations via tissue compatability studies following, e. g., skin grafts.

Since the cytonucleoproteins (see Section III A 2) have a unique behavior that can be clearly characterized, they can serve in specificity studies. If an *Amoeba proteus* nucleus labeled with incorporated radioactive amino acids is grafted into the closely related *A. discoides*[3], we find that the labeled cytonucleoproteins (of *A. proteus*) continue to behave in the manner we have already described; there is no apparent difference in cytonucleo-

---

[3] It is difficult to conclude that *A. discoides* differs from *A. proteus* sufficiently to be called a separate species (Kates and Goldstein 1963).

protein behavior from that observed when an *A. proteus* to *A. proteus* operation is performed. On the other hand if a radioactive protein-labeled *A. proteus* nucleus is grafted into *Chaos chaos,* a species that is recognizably different from *A. proteus* even to the uninitiated, the autoradiographic picture observed after the equilibrium distribution of labeled protein is reached is noticeably different from that of the former distribution. In the *A. proteus*—*C. chaos* case, although radioactivity appears in relatively high concentration in the host cell nuclei, there is a substantial amount of radioactivity detectable in the cytoplasm (Byers et al. 1963). Thus, there appears to be less compatibility between *A. proteus* cytonucleoproteins and *C. chaos* than between *A. proteus* cytonucleoproteins and *A. discoides.* This is, of course, no great surprise but does suggest the possibility of investigating a number of more subtle interspecies relationships.

What is the cause of the pronounced labeling of *C. chaos* cytoplasm after the introduction of radioactive *A. proteus* cytonucleoproteins? Does it represent breakdown of *A. proteus* proteins and random reincorporation of the products? Does it mean that there are several classes of cytonucleoproteins and that only some are closely related in the two species? Or does it mean that a higher proportion of cytonucleoproteins in *C. chaos* is present in the cytoplasm at any instant? Obviously we should like to do more to clarify the relations between species with respect to these macromolecules and perhaps discover new interactions of these components as well as other components not yet studied.

IV. Conclusion

This article represents an attempt to show some of the potential value in the combined use of nuclear transplantation and isotope techniques to solve some contemporary cell biology problems. The worth of these techniques in combination has been demonstrated on several levels: more direct answers to some problems can be obtained than is true for other methods; some phenomena could not be studied or discovered by other current techniques; and it is possible to critically test the validity of some commonly used procedures. Since our techniques are difficult to apply, we hope that the findings so obtained will provide insights for investigations by more standard procedures that are simpler or that can be performed on a "macro" scale.

We have illustrated here only the simplest experimental designs; no consistent effort has been made to alter the physiological activity of the experimental cells. There are, obviously, many possibilities for manipulating the extracellular and intracellular environments and noting the effects on the few phenomena described here. Thus far only some work on the effect of extracellular radiation on the movement of RNA (Iverson 1962) and limited studies on the effect of temperature on cytonucleoprotein movement (Byers et al. 1963) have been carried out. The effects of all sorts of chemical factors, genetic factors, metabolic states, etc. remain to be explored. Such studies are necessary because (aside from some knowledge of the behavior of RNA) little is known about the physiological role of the molecules that have been investigated.

It is the author's not very startling suspicion that the techniques heralded here will prove their greatest future value in studies on the interaction between cytoplasm and nucleus. Little is known, e. g., about how the cytoplasm influences gene activity, yet it is increasingly evident that this is an area of cell behavior about which we should like to know a great deal more. The behavior of tagged molecules following nuclear transplantation promises to afford some important clues to many kinds of nucleocytoplasmic interactions.

References

ALLFREY, V., 1959: In "The Cell" (J. BRACHET and A. E. MIRSKY, eds.), I, p. 193. Academic Press, New York.
BRACHET, J., 1957: Biochemical Cytology. Academic Press, New York.
BRENNER. S., F. JACOB, and M. MESELSON, 1961: Nature 190, 576.
BYERS, T. J., D. B. PLATT, and L. GOLDSTEIN, 1963: J. Cell Biol. 19, 453, 467.
CHADWICK, A., 1961: Exper. Cell Res. 25, 131.
COMMANDON, J., and P. DE FONBRUNE, 1939: Comp. rend. soc. biol. 130, 740.
DANIELLI, J. F., I. J. LORCH, M. J. LORD, and E. G. WILSON. 1955: Nature 176, 1114.
DE FONBRUNE. P., 1949: Technique de Micromanipulation. Masson, Paris.
DOUNCE, A. L., 1953: Nature 172. 541.
GOLDSTEIN. L.. 1958: Exper. Cell Res. 15. 635.
— 1964: In "Cytology and Cell Physiology" (G. H. BOURNE, ed.) 3rd Edition, Academic Press. New York.
— 1963 a: Symp. Internat. Soc. Cell Biol. (E. J. C. HARRIS, ed.) II, 129.
— and J. MICOU, 1959: J. Biophys. Biochem. Cytol. 6, 1.
— and W. PLAUT, 1955: Proc. Nat. Acad. Sci. U.S. 41, 874.
— J. MICOU, and T. T. CROCKER, 1960: Biochem. Biophys. Acta 45, 82.
HARRIS, H., 1959: Biochem. J. 73, 362.
HOGEBOOM. G. H., and W. C. SCHNEIDER, 1952: J. Biol. Chem. 197, 611.
HOLTER. H., 1943: Compt.-rend. Lab. Carlsberg, Sér. chim. 24, 399.
IVERSON. R. M.. 1962: Exper. Cell Res. 27. 125.
KATES, J. R., and L. GOLDSTEIN, 1964: J. Protozoology (in press).
KROEGER, H., J. JACOBS, and J. L. SIRLIN, 1963: Exper. Cell Res. 31, 416.
MIRSKY, A. E., and S. OSAWA, 1961: In "The Cell" (J. BRACHET and A. E. MIRSKY, eds.) II, p. 677. Academic Press, New York.
PERRY, R. P., 1962: Proc. Nat. Acad. Sci. U. S. 48, 2179.
PRESCOTT, D. M., 1960: Ann. Rev. Physiol. 22, 17.
— 1961: In: "Biological Structure and Function", (T. W. GOODWIN and O. LINDBERG, eds.) II. p. 257. Academic Press, New York.
— 1963: Symp. Internat. Soc. Cell Biol. (R. J. C. HARRIS, ed.) II, 111.
RICH, A., and J. D. WATSON, 1954: Proc. Nat. Acad. Sci. U. S. 40, 759.
ROODYN, D. B., 1959: Internat. Rev. Cytol. 8. 279.
WALKER. P. M. B., and B. M. RICHARDS, 1959: In: "The Cell" (J. BRACHET and A. E. MIRSKY, eds.) I, p. 91. Academic Press, New York.
WOODS, P. S., and J. H. TAYLOR. 1959: Lab. Invest. 8, 309.
ZALOKAR, M., 1959: Nature 183, 1330.

Subject Index

Fortsetzung von der 4. Umschlagseite

Le chondriome de la cellule végétale: morphologie du chondriome. Par Pierre Dangeard, Bordeaux. Avec 23 figures. IV, 35 pages. — **Die Sphärosomen der Pflanzenzelle.** Von E. S. Perner, Münster, Westfalen. Mit 25 Textabbildungen. 71 Seiten. Gr.-8°. 1958. Band III. Cytoplasma-Organellen. A. Chondriosomen, Mikrosomen, Sphärosomen. 1, 2.

S 252.—, DM 40.—, sfr. 43.—, $ 10.—

Chemistry of Viruses. By C. A. Knight, Berkeley (California). With 27 figures. IV, 177 pages. Gr.-8°. 1963. Band IV. Virus. 2.

S 303.—, DM 48.—, sfr. 51.60, $ 12.—

The Multiplication of Viruses. By S. E. Luria, Urbana, Illinois. IV, 63 pages. — **Virus inclusions in Plant Cells.** By Kenneth M. Smith, Cambridge. With 5 plates. 16 pages. — **Virus Inclusions in Insect Cells.** By Kenneth M. Smith, Cambridge. With 16 figures. 25 pages. — **Antibiotika erzeugende virus-ähnliche Faktoren in Bakterien.** Von Pierre Fredericq, Lüttich. 14 Seiten. Gr.-8°. 1958. Band IV. Virus. 3, 4a, 4b, 5.

S 268.—, DM 42.50, sfr. 45.70, $ 10.65

Strukturtypen der Ruhekerne von Pflanzen und Tieren. Von Elisabeth Tschermak-Woess, Wien. Mit 91 Textabbildungen (427 Einzelbildern). IV, 158 Seiten. Gr.-8°. 1963. Band V. Karyoplasma (Nucleus). 1.

S 353.—, DM 56.—, sfr. 60.20, $ 14.—

Riesenchromosomen. Von Wolfgang Beermann, Tübingen. Mit 113 Textabbildungen. IV, 161 Seiten. Gr.-8°. 1962. Band VI. Kern- und Zellteilung. D.

S 312.—, DM 49.50, sfr. 53.20, $ 12.40

Die Amitose der tierischen und menschlichen Zelle. Von Otto Bucher, Lausanne. Mit 56 Textabbildungen. IV, 159 Seiten. Gr.-8°. 1959. Band VI. Kern- und Zellteilung. E. Amitose. 1.

S 426.—, DM 67.50, sfr. 72.60, $ 16.90

Les altérations de la méiose chez les animaux parthénogénétiques. Par Marguerite Narbel-Hofstetter, Dr ès Sc., Privat-docent, Laboratoire de Zoologie et d'Anatomie comparée de l'Université de Lausanne. In französischer Sprache. Mit 112 Textabbildungen (686 Einzelbilder). IV, 163 Seiten. Gr.-8°. 1964. Band VI. Kern- und Zellteilung. F. Die Chromosomen in der Meiose. 2.

S 397.—, DM 63.—, sfr. 67.70, $ 15.75

Différenciation des cellules sexuelles et Fécondation chez les Phanérogames. Par Bernard Vazart, Bondy (Seine). Avec 54 figures. IV, 158 pages. Gr.-8°. 1958. Band VII. Befruchtung und Kernverschmelzung. 3a.

S 378.—, DM 60.—, sfr. 64.50, $ 15.—

Différenciation des cellules sexuelles et Fécondation chez les Cryptogames. Par Bernard Vazart, Bondy (Seine). Avec 122 figures. IV, 363 pages. Gr.-8°. 1963. Band VII. Befruchtung und Kernverschmelzung. 3b.

S 706.—, DM 112.—, sfr. 120.40, $ 28.—

Protoplasmic Streaming. By Noburô Kamiya, Osaka, Japan. With 82 figures. IV, 199 pages. Gr.-8°. 1959. Band VIII. Physiologie des Protoplasmas. 3. Motilität. a.

S 472.—, DM 75.—, sfr. 80.60, $ 18.75

Frost, Drought, and Heat Resistance. By J. Levitt, Columbia, Missouri. With 29 figures. IV, 87 pages. Gr.-8°. 1958. Band VIII. Physiologie des Protoplasmas. 6.

S 220.—, DM 35.—, sfr. 37.60, $ 8.75

Polarität und inäquale Teilung des pflanzlichen Protoplasten. Von Erwin Bünning, Tübingen. Mit 72 Textabbildungen. IV, 86 Seiten. Gr.-8°. 1958. Band VIII. Physiologie des Protoplasmas. 9. Polarität. a.

S 220.—, DM 35.—, sfr. 37.60, $ 8.75

Effets biologiques des radiations. Aspects biochimiques. Par Maurice Errera, Bruxelles. Avec 27 figures. IV, 241 pages. Gr.-8°. 1957. Band X. Pathologie des Protoplasmas. 3.

S 488.—, DM 77.50, sfr. 83.30, $ 19.40

Morphology and Physiology of Plant Tumors. By Armin C. Braun and Tom Stonier, New York, N. Y. With 7 figures. IV, 93 pages. Gr.-8°. 1958. Band X. Pathologie des Protoplasmas. 5a.

S 206.—, DM 32.50, sfr. 34.90, $ 8.15

Protoplasmatische Ökologie der Pflanzen. Wasser und Temperatur. Von Richard Biebl, Wien. Mit 92 Textabbildungen. IV, 344 Seiten. Gr.-8°. 1962. Band XII. Protoplasmatische Ökologie der Pflanzen. 1.

S 618.—, DM 98.—, sfr. 105.40, $ 24.50